MEASURING THE MARGINAL SOCIAL COST OF TRANSPORT

RESEARCH IN TRANSPORTATION ECONOMICS

Series Editor: Martin Dresner

RESEARCH IN TRANSPORTATION ECONOMICS VOLUME 14

MEASURING THE MARGINAL SOCIAL COST OF TRANSPORT

EDITED BY

CHRISTOPHER NASH

Institute for Transport Studies, University of Leeds, U.K.

BRYAN MATTHEWS

Institute for Transport Studies, University of Leeds, U.K.

ELSEVIER
JAI

Amsterdam – Boston – Heidelberg – London – New York – Oxford
Paris – San Diego – San Francisco – Singapore – Sydney – Tokyo

ELSEVIER B.V.
Radarweg 29
P.O. Box 211
1000 AE Amsterdam,
The Netherlands

ELSEVIER Inc.
525 B Street, Suite 1900
San Diego
CA 92101-4495
USA

ELSEVIER Ltd
The Boulevard, Langford Lane, Kidlington
Oxford OX5 1GB
UK

ELSEVIER Ltd
84 Theobalds Road
London
WC1X 8RR
UK

© 2005 Elsevier Ltd. All rights reserved.

This work is protected under copyright by Elsevier Ltd, and the following terms and conditions apply to its use:

Photocopying
Single photocopies of single chapters may be made for personal use as allowed by national copyright laws. Permission of the Publisher and payment of a fee is required for all other photocopying, including multiple or systematic copying, copying for advertising or promotional purposes, resale, and all forms of document delivery. Special rates are available for educational institutions that wish to make photocopies for non-profit educational classroom use.

Permissions may be sought directly from Elsevier's Rights Department in Oxford, UK: phone (+44) 1865 843830, fax (+44) 1865 853333, e-mail: permissions@elsevier.com. Requests may also be completed on-line via the Elsevier homepage (http://www.elsevier.com/locate/permissions).

In the USA, users may clear permissions and make payments through the Copyright Clearance Center, Inc., 222 Rosewood Drive, Danvers, MA 01923, USA; phone: (+1) (978) 7508400, fax: (+1) (978) 7504744, and in the UK through the Copyright Licensing Agency Rapid Clearance Service (CLARCS), 90 Tottenham Court Road, London W1P 0LP, UK; phone: (+44) 20 7631 5555; fax: (+44) 20 7631 5500. Other countries may have a local reprographic rights agency for payments.

Derivative Works
Tables of contents may be reproduced for internal circulation, but permission of the Publisher is required for external resale or distribution of such material. Permission of the Publisher is required for all other derivative works, including compilations and translations.

Electronic Storage or Usage
Permission of the Publisher is required to store or use electronically any material contained in this work, including any chapter or part of a chapter.

Except as outlined above, no part of this work may be reproduced, stored in a retrieval system or transmitted in any form or by any means, electronic, mechanical, photocopying, recording or otherwise, without prior written permission of the Publisher.

Address permissions requests to: Elsevier's Rights Department, at the fax and e-mail addresses noted above.

Notice
No responsibility is assumed by the Publisher for any injury and/or damage to persons or property as a matter of products liability, negligence or otherwise, or from any use or operation of any methods, products, instructions or ideas contained in the material herein. Because of rapid advances in the medical sciences, in particular, independent verification of diagnoses and drug dosages should be made.

First edition 2005

British Library Cataloguing in Publication Data
A catalogue record is available from the British Library.

ISBN-10: 0-7623-1006-5
ISBN-13: 978-0-7623-1006-7
ISSN: 0739-8859 (Series)

∞ The paper used in this publication meets the requirements of ANSI/NISO Z39.48-1992 (Permanence of Paper).

Transferred to digital printing 2006
Printed and bound by CPI Antony Rowe, Eastbourne

Working together to grow libraries in developing countries

www.elsevier.com | www.bookaid.org | www.sabre.org

ELSEVIER BOOK AID International Sabre Foundation

CONTENTS

LIST OF CONTRIBUTORS

Ofelia Betancor	Universidad de Las Palmas de Gran Canaria, Spain
Peter Bickel	Universität Stuttgart, Germany
Ian Black	Cranfield University, U.K.
Miguel Carmona	TIS, Lisboa, Portugal
Corina Certan	NEI Transport, Netherlands
Claus Doll	Fraunhofer-Institute for Systems and Innovation Research, Germany
Rainer Friedrich	Universität Stuttgart, Germany
Jan Owen Jansson	Universitet I Linkoping, Sweden
Gunnar Lindberg	TEK/VTI, Sweden
Heike Link	German Institute for Economic Research (DIW Berlin), Germany
Rosário Macário	TIS, Lisboa, Portugal
Bryan Matthews	University of Leeds (ITS), U.K.
Inge Mayeres	Federal Planning Bureau, Belgium
Chris Nash	University of Leeds (ITS), U.K.
Jan-Eric Nilsson	TEK/VTI, Sweden
Stef Proost	Katholieke Universiteit Leuven, Belgium
Emile Quinet	CERAS, France

Andrea Ricci	ISIS, Italy
Stephan Schmid	DLR, Germany, formerly Universität Stuttgart
Marten van den Bossche	NEI Transport, Netherlands
Kurt Van Dender	University of California, Irvine, USA
Simme Veldman	NEI Transport, Netherlands

ACKNOWLEDGEMENTS

The editors would first like to thank all of the contributing authors for their efforts and cooperation in drawing this book together. Without these valuable contributions this book would not have been possible. More generally, we would like to record our thanks to colleagues who have collaborated with us on the projects on which this book is based; most notably, the UNITE project, but also the PETS, CAPRI, IMPRINT-EUROPE and MC-ICAM projects. Furthermore, we would like to acknowledge the Directorate General for Energy and Transport (DG TREN) for having helped stimulate this area of research through its pursuit of pricing policy reforms and its funding of research in this area. In particular, we are grateful for the efforts of the project officers at DG TREN who have overseen the above mentioned research projects, Catharina Sikow-Magny and Christophe Deblanc. Finally, a special word of thanks to Julie Whitham (ITS, University of Leeds) who, having undertaken the task of turning huge volumes of text from all over Europe into the set of UNITE deliverable with her usual blend of efficiency and good humour, then did the same for this book.

FOREWORD

Transport taxation and pricing is a complex issue. It has been known ever since the French engineer Jules Dupuit in the 19th century wrote that transport taxes and prices should reflect the marginal cost of using the transport system, such as congestion and other external costs. Yet, the practical implementation of marginal cost pricing has been rare in practice – except for the recent London congestion charging scheme and for some elements of heavy good vehicle tolling systems in central Europe. The existing charging regimes typically rely on taxes that only remotely, if at all, reflect the underlying marginal costs, and revenue-raising objectives.

The European Commission has advocated the reform of transport pricing for over a decade now. The Commission's green paper *Towards Fair and Efficient Pricing in Transport* (1996) launched discussion on pricing transport according to marginal cost and on recovery of fixed investment cost. The white paper *Fair Payment for Infrastructure Use* (1998) took the approach a step further and presented a gradual implementation path for all transport modes, which was taken up in the white paper *European Transport Policy 2010: Time to Decide* (2001). According to this policy, transport taxes and prices should vary according to infrastructure damage, degree of congestion, accident risk and environmental nuisances. While this would change the structure of taxes and charges sometimes dramatically, the overall charge levels would not necessarily change.

In parallel, intensive research on transport pricing has been carried out both at the national and community levels. Without this, progress in policy would not have been possible and the endless debates (on what pricing instruments are most efficient and what the appropriate price levels should be in different contexts etc.) would have continued. The most pertinent research questions have been the following:

How to operationalise marginal cost pricing principles in practice? Which charging instruments (tolls, parking charges, fuel taxes, etc.) lead to efficient outcomes with least cost and distortion?

What are the relevant external and internal costs of transport? How can they be measured and valued in different contexts, e.g. in interurban corridors and urban areas? What are the socio-economic impacts of implementing efficient pricing in different regions and for different citizen groups in Europe? When and how to design mitigation or compensatory measures?

What is the outcome of efficient pricing in relation to cost recovery, financing and public–private partnerships in trans-European networks and urban public transport systems?

Which measures should be taken to increase public and political acceptability of pricing measures at system and city levels and in particular for roads?

Research activities under the Community Framework Programmes have looked into most of the above issues. While the earlier projects under the 4th Framework Programme have addressed more theoretical issues, such as the application of the marginal cost principle and the determination of optimal charging instruments, the focus in the 5th and 6th Framework Programmes has clearly shifted from theory to practical implementation of efficient pricing. The projects have focussed on developing tools to estimate marginal costs, on assessing the impacts of efficient pricing on demand, modal shares and economic growth. Also, demonstration projects in several European cities have been carried out to test the practical feasibility and acceptability of different pricing schemes and other transport policy measures, such as traffic management, park and ride. In parallel, to ensure wide dissemination of the results of the Framework Programme pricing research but also to allow for a debate among practitioners, several networking actions have also been funded.

The UNITE project, funded under the 5th Framework Programme in 2000–2002, had three interlinked objectives:

Development of pilot *transport accounts* that give a clear and transparent overview of the economic and financial flows of the transport system and of the individual modes. The accounts were compiled for all the 15 EU Member States plus Switzerland, Hungary and Estonia as well as for an urban area.

Estimation of *marginal costs* of using the transport system in different local and traffic contexts. The project undertook some 30 case studies covering all relevant external cost categories, modes, traffic contexts and Member States.

Integration of the accounts and marginal cost approaches and recommendations on how to apply cost figures from one location/case study in another place.

The results of the UNITE research were presented in an international conference in Leuven, Belgium in summer 2002 bringing together some

70 researchers, practitioners and policy makers from all over Europe and beyond. There was a lot of discussion on several key issues: the difference between the average costs derived from the accounts and the marginal costs from the case studies; on the marginal cost values calculated for road accident risk and for the environmental performance of different aircraft types or for scarcity in rail, as well as on the difficulty to generalise results from one place to another. On one issue there was consensus among all the participants: UNITE had succeeded in providing the state of the art in the methodology on transport pricing and it had compiled a valuable database on costs. These have since been taken up by several research projects on transport pricing and contributed to the development of pricing policies in several Member States.

UNITE has been one of the most challenging projects on transport pricing that I have managed as a Commission scientific officer. This is for many reasons. The research objectives of the project were ambitious from the methodological and empirical points of view. The coverage of the work, both in terms of costs to be looked at and geographically, was huge. In addition, the UNITE "dream team", as one of the Consortium members once called it, comprising 19 leading European institutes in pricing research, called for a major effort in co-ordination. Despite this or maybe because of this, UNITE was also one of the most pleasant projects that I have managed for the Commission. I would therefore like to express my thanks to the whole "dream team" of UNITE and in particular to the project coordinator Professor Chris Nash, ITS, Leeds for the quality of the work done and the excellent cooperation.

Catharina Sikow-Magny
European Commission

TRANSPORT PRICING POLICY AND THE RESEARCH AGENDA

Chris Nash and Bryan Matthews

1. BACKGROUND

The use of transport systems imposes a range of costs, both within the transport system and elsewhere in the economy. When a lorry, for example, enters the traffic flow, it imposes some degree of damage to the road surface; it adds to the number of vehicles on the road and, according to standard speed flow relationships, will have some impact on overall traffic speeds; it emits from its exhaust pipe fumes, which contribute to air pollution and to global warming; it generates noise from its engine and from the rub of its wheels on the road; and it contributes in some way to the risk of a road traffic accident occurring.

The imposition of these costs of transport use, and in particular the failure to reflect them within decisions about whether, when, where and how to travel, gives rise to a range of problems that impact on the transport system and have wider repercussions for the environment and economy. Hence, the combined effects of damage being imposed on the infrastructure have contributed to that infrastructure sometimes being in a poor state of repair. The combined effects of greater numbers of vehicles – be they cars, lorries, trains or aircraft – has resulted in roads, railways and skies becoming increasingly congested at certain times and places, resulting in journey-time delays and unreliability, over-crowding and scheduling problems. According to the UNITE project, it is estimated that, in Europe, costs associated with road congestion – roads being the most congested of the transport modes – amount to 70 billion Euro, approximately 1% of its overall GDP

Measuring the Marginal Social Cost of Transport
Research in Transportation Economics, Volume 14, 1–18
Copyright © 2005 by Elsevier Ltd.
All rights of reproduction in any form reserved
ISSN: 0739-8859/doi:10.1016/S0739-8859(05)14001-3

(Nash et al., 2003). In addition to these problems faced by the transport system and which have direct effects on transport-related activities, the transport system is a source of problems that have wider implications for the whole of society. Environmental pollution and external accident costs of road transport – roads again being the most polluting and most dangerous of the transport modes – taken together, were found to be at least 122 billion Euro, approximately 1.6% of Europe's GDP (Nash et al., 2003). These costs considerably exceed the costs of providing and maintaining infrastructure, suggesting that they should be central to transport pricing decisions. The failure to reflect the costs imposed by users of the transport system in their individual travel decisions has the effect of imposing costs on others who are engaged in travel, on governments that intervene in an effort to alleviate these problems and on the whole of society.

In recent years, the response of governments to these problems has increasingly been to place greater weight on economic instruments as a way to tackle them, by giving transport users appropriate incentives to modify their behaviour. Probably the most explicit and co-ordinated expression of this policy has come from the European Commission, the executive arm of the European Union, in the development and implementation of its transport pricing policy over the past 10 years. Starting with the Commission's Green Paper 'Towards Fair and Efficient Pricing in Transport' (CEC, 1995a), and continuing with the White Paper 'Fair Payment for Infrastructure Use' (CEC, 1998) and the new Common Transport Policy White Paper (CEC, 2001), there is a strong emphasis on pricing policy to reflect the full social costs of transport use. However, pricing is becoming an increasingly prominent feature of transport policy, in particular for roads, throughout the world. Perhaps the most notable developments have occurred in Singapore where there has been a system of road pricing since 1975, in Norway where several urban road pricing schemes have been introduced since 1986, in the U.S.A. where several road pricing schemes have been introduced as part of the Value Pricing Programme, in Switzerland where a heavy vehicle fee was introduced in 2001 and most recently with the city centre area charge in London in 2003.

The publication by the European Commission of its Green Paper 'Towards fair and efficient pricing in transport' (CEC, 1995a) represented a major step forward in transport pricing policy development at the European level. Whereas previous discussion of EC pricing policy had emphasised maintenance and operating costs, this paper recognised the importance of pricing to reflect external costs.

The policy was taken further in the White Paper on 'Fair payment for infrastructure use' in 1998 (CEC, 1998). The latter put a clear case for marginal cost pricing, whilst recognising that the movement towards this target would need to be phased over a number of years, and that second best measures to achieve desired levels of cost recovery would continue to be necessary. A number of mode-specific pricing policy developments have either stemmed from these papers or have emerged in parallel with them.

At the same time, the Commission has funded a whole range of research projects into transport pricing, tackling the many different issues that arise in implementing this policy. Initially, a key issue was research on how to apply the principles of marginal social cost pricing within the context of transport, in particular transport infrastructure. A major research effort has subsequently focused on how to measure the marginal social costs of transport and, using the methods developed, on providing estimates of marginal social costs for use in price-setting. There has also been an important body of research into the impacts that would arise if marginal social cost pricing in transport were implemented. More recently, research efforts have focused on the question of how to implement the policy, as well as on resolving the remaining uncertainties emerging from the research on measurement and impacts. In addition to this research, considerable effort has been put into communicating the policy, disseminating research findings and building consensus amongst stakeholders.

This opening chapter seeks to provide the context for the subsequent chapters by setting out how transport pricing policy has developed over recent years and how research needs have emerged and been addressed. We first elaborate how the basic principles of marginal social cost pricing can be applied within the transport sector. We then summarise the comprehensive critique of the theory and its application to transport, and the various responses to the particular criticisms and make the case for marginal social cost 'based' pricing. Having addressed the theoretical arguments, we then seek to summarise policy developments within the European Union. Next, we consider the research needs that have emerged from this policy development and how they have been addressed, in particular focusing on the major research programmes of the European Commission. Finally, we provide an overview of the subsequent chapters of the book.

2. APPLYING MARGINAL SOCIAL COST PRICING THEORY TO TRANSPORT

A number of European research projects, including in particular PETS, TRENEN and AFFORD, sought to clarify the approach advocated by the European Commission to transport pricing. Essentially, this approach is that known to economists as short-run marginal cost pricing, whereby prices are set to reflect the additional costs to society associated with an additional kilometre travelled or an additional trip made, given that the capacity of the transport network is held constant.

When car users, rail operators or the operators of other vehicles decide to travel additional kilometres or to make additional trips, they impose additional costs on themselves, on the infrastructure-provider, on other users and on the rest of society. Costs to other users and to the rest of society are referred to as 'external' costs. External costs or benefits arise when the social or economic activities of one agent have an impact on the welfare of another agent, without that impact having been taken into account by the first agent. If monetary values can be placed upon externalities, then they can be incorporated into the pricing mechanism by means of direct charges or subsidies; in this way they will then be taken into account by all economic agents.

Prices that reflect the additional infrastructure and external costs will act as signals to travellers about the 'social' costs associated with their additional travel. They will then base their demand decisions – whether, where, when, how and how far to travel – upon these price signals. In fact, prices fulfil several functions in parallel. In addition to acting as cost signals, the price mechanism is the best way to ensure that a limited supply of a good is made accessible to those who value it most. By raising prices until the total demand equals the available quantity, the consumers with the highest willingness to pay for the good receive the good. Also, in competitive markets firms will only succeed if their prices are kept as low as possible; otherwise their competitors will take their markets. In this way the price mechanism provides all producers with incentives to develop cost-reducing production techniques.

What the principle of short-run marginal cost pricing translates into, in terms of infrastructure charging, is a need to measure three components of cost for the addition of extra traffic to the existing infrastructure. The first is the cost imposed by additional use on the infrastructure provider. This comprises additional maintenance and renewals costs plus any additional operating costs. The second component is the marginal cost imposed on other infrastructure users, in terms of delays, congestion, accidents and opportunity costs (perhaps more commonly referred to as scarcity costs), on those modes where there is a physical limit and once all the slots are taken no one else can get one. The third element is the cost imposed outside the transport system and that is mainly environmental cost, but some elements of other costs such as accidents, for instance, where these are borne in part by the police or health service and not recovered from users, may enter there. Note that it is only environmental costs related to vehicle use that are relevant in this context; environmental costs related to infrastructure provision are relevant for investment decisions but not (directly) for pricing policy. (There is an indirect effect. If environmental costs limit infrastructure provision, for instance, in cities or in environmentally sensitive areas, this will raise congestion and scarcity costs and thus raise price).

The same sort of approach may be taken to scheduled transport services. In the case of private transport, if the infrastructure prices are right, essentially, the problem of efficient use of the system is solved. But with scheduled public transport services and with freight transport services that is not so. Or at least it is not so unless there is a fully competitive environment so that it can be left up to the market to determine what is produced. In practice, we rarely have that and there are various cost characteristics, of scheduled transport in particular, which make that difficult and unlikely. For instance, when traffic is added to public transport systems, either this raises load factors or leads to operation of larger vehicles or longer trains, in which case the marginal cost to the operators is very low, or services are increased, in which case there is a benefit to existing users from a better service as traffic rises. In other words, for the marginal social cost of operating scheduled transport services, there is again a mix of costs to the supplier, to the users and to society at large. But the cost to other users is typically negative because extra traffic leads to an improvement in the service (Mohring, 1972). This means that there is very often an a priori case for subsidising scheduled transport services in order to implement pricing policies that do not cover full cost. In the absence of efficient provision of the scheduled transport services themselves there is no guarantee that simply getting the infrastructure charging right will even improve resource allocation let alone solve the problem. The Commission has been concerned mainly with infrastructure charging because of its concern with the terms of competition between different users of the infrastructure as it promotes open access and competitive markets for all modes of transport. But in doing so, it has given less attention to a very important aspect of transport pricing: the point is that for scheduled transport services, it is the final price to the consumer that determines its competitive position with respect to other modes.

3. A CRITIQUE OF MARGINAL COST-BASED PRICING

The most thoroughly argued critique of the approach published to date is that of Rothengatter (2003). According to Rothengatter, the Commission has essentially adopted the marginal social cost pricing approach of neo-classical economic textbooks. This discussion of Rothengatter's critique follows that in Nash (2003).

As Rothengatter rightly reminds us, there are numerous reasons why the simple textbook approach may not be optimal in practice. These may be summarised as follows:

(a) measurement is complex
(b) equity is ignored
(c) dynamic effects, including investment decisions and technology choice, are ignored
(d) financing issues are ignored
(e) institutional issues are ignored
(f) price distortions elsewhere in the economy are ignored
(g) implementing marginal social cost pricing may involve substantial administrative costs, which may not always be justified by the benefits it brings.

All of these criticisms are well established in the literature and can hardly be denied. Do they imply, as Rothengatter asserts, that there is no need for a uniform pricing system and that a real-world pricing system cannot be based on an abstract economic orthodoxy? We will consider these criticisms one by one.

(a) That measurement of short-run marginal social cost is complex is undeniable. A recent review (Lindberg, 2002) concludes that even those elements of cost that have been long studied, such as infrastructure maintenance and renewals and congestion cost, are not without considerable uncertainty as to the true marginal social cost. Scarcity costs, which arise on those modes where use of the infrastructure is scheduled and the presence of operators filling all the slots make it impossible for anyone else to get access to the infrastructure at the time in question, are little researched. Whilst enormous progress has been made on the measurement and valuation of environmental costs and external accident costs, these too are still subject to big uncertainties.

However, Lindberg concluded that research within a number of European projects is rapidly reducing this uncertainty, and that 'the use of proper theory and modern methods will lead to a convergence also of the more difficult marginal cost categories in the near future'. In other words, there is no reason for measurement problems to hold up moves towards marginal social cost pricing. In any event it is hard to argue that, were marginal social cost the right concept to use in pricing, measuring something else instead of using the best estimate possible would be a sensible approach.

(b) The issue of equity is very controversial. If the Green Paper was not totally clear on what constituted efficient pricing, it was much less clear on what was fair. Fairness seemed to be based on some version of the polluter pays principle, whereby the user paid the costs they imposed. But this principle might be applied at the margin, requiring that the individual consumer should pay at least the additional costs they impose, or it might be applied at a more aggregate level, for example, requiring that all users of a particular mode of transport should collectively bear the total costs this mode of transport imposes. Rather remarkably, by the time the Commission issued a White Paper on Infrastructure Charges (CEC, 1998), which firmly espoused the marginal cost pricing principle, it was entitled Fair Payment for Infrastructure Use, the word 'efficient' having been dropped rather than the word 'fair'!

It may be argued that a more appropriate way of dealing with equity approaches the problem in a totally different way, as concerning the overall welfare of each individual or group of individuals in society, rather than the effects of a single policy or project (Mayeres & Proost, 2003). Thus if someone is poor, or disadvantaged in some other way – for instance through lack of affordable transport – then policy needs to seek to offset this disadvantage; if someone is clearly well favoured, then they represent a prime target for the levying of additional charges. Of course in an ideal world this would all be done through the taxation and income supplementation system and pricing of transport services could then ignore equity as an issue. But given that it is not politically or practically possible to deal with all these issues in this way, they do need to be taken into account in pricing decisions. It is necessary therefore to know who will gain and who will lose by these decisions. Thus, for instance, it may be seen as fairer to expand subsidies to bus services than to air services or high speed trains, even though on pure efficiency grounds the argument may go the other way. But the appropriate approach to dealing with equity concerns of this type is to use distributive weighting systems, which still use marginal social cost pricing as a starting point, but adjust prices in accordance with the distributive characteristics of the goods concerned (Feldstein, 1972). There is no reason to suppose that this rule will in general lead to full cost recovery on any particular mode or facility.

(c) The dynamic concerns relate particularly to the fact that short-run marginal social cost pricing totally ignores the capital costs of expanding the system. There is an alternative in the textbooks, known as long-run marginal cost pricing, that charges not the costs of adding extra traffic to the existing infrastructure, but the costs imposed by the extra traffic when the infrastructure is optimally adjusted to the new traffic level. This will therefore include some marginal capital costs, but compared with short-run marginal cost the congestion cost, and possibly some of the other external costs, will be reduced. It is easy to show that if capacity is optimally adjusted, then it is expanded up to the point where the extra capital costs of expanding capacity just equal the reduced other costs, so that at the margin short- and long-run marginal costs are equal. There is therefore only an issue between the two as a basis for charging when capacity is not always optimally adjusted to demand. Resolution of this issue depends on the relative speeds of adjustment of capacity to demand and of consumers to price. It is clear that infrastructure capacity adjusts slowly to demand. If consumers adjusted instantaneously to price, then short-run marginal social cost pricing would clearly mean that at all points in time demand would be optimally adjusted to capacity. There remains an issue of the incentives for the appropriate adjustment of capacity over time, which we will return to under-institutional issues. Given that there is in fact a lag in adjustment of demand to price, particularly for commuters and for freight traffic – where the adjustment may involve relocation – it is not necessarily the case that charging short-run marginal social cost will always achieve optimal adjustment of demand to capacity, and there is clearly a case for at least smoothing the adjustment of price as capacity changes to give signals as to the longer term level of marginal social cost.

If we moved entirely to long-run marginal cost as the basis for charging, then the case for charging higher prices in urban areas would be based on the high cost of expanding capacity in such locations rather than on existing levels of congestion. But in practice, most advocates of long-run marginal cost pricing would accept that in some locations, particularly in urban areas, the costs of expanding capacity are so high and the political problems so great that it is inappropriate to price on this basis – it is more important to get efficient use of existing capacity by pricing at short-run marginal cost. It is only in locations, perhaps on the public transport and inter-urban road network, where infrastructure capacity expansion is actually likely to take place, that there is a case for long-run marginal cost pricing. But there are many other problems, such as indivisibilities, which mean that in practice, the measurement of short-run marginal social cost, but allowing for a different level of infrastructure capacity, may still be the best approach. In short, dynamic effects may mean modification of the simple short-run marginal social cost pricing rule, but the concept of marginal cost is still fundamental to the derivation of an efficient alternative.

(d) The financing and institutional issues reflect the most common objection to marginal cost pricing – that it does not recover total cost. However, this conclusion may depend on the level of aggregation at which the comparison of costs and revenue is made. At the aggregate level, pricing to recover total cost typically implies big increases in rail and other public transport charges, together with reductions in road taxation, whereas marginal cost pricing often implies the reverse. Most transport infrastructure is subject to increasing returns to scale, which means if capacity is anything like optimal, then marginal cost pricing will not recover the total cost of the system. On the other hand, the cost of land and property acquisition limit expansion, particularly in urban areas, so efficient supply of capacity in urban areas will typically still involve significant congestion and consequently high charges for the use of infrastructure. So the likelihood is that marginal cost pricing will imply substantial surpluses of revenue over costs on urban roads, whereas on rural roads and much public transport, there will be deficits. Nevertheless, there remain two important questions. One is whether the whole package of effects satisfies government budget constraints. Does it provide enough finance, or are there other means of supplementing it if necessary? There is some evidence that typically surpluses in urban areas are so big that budget constraints are not a problem (Roy, 2002). But then there is a second issue, and that is whether a system whereby urban road users greatly cross-subsidise rural road users and public transport users is perceived as acceptable, both in terms of equity and in terms of influence on locational choice.

If for any of these reasons the budgetary outcome of marginal social cost pricing is seen as unacceptable, then of course we have to depart from marginal cost pricing, but that does not mean we throw the principles away. There are well-established rules for supplementing the revenue raised in a way that is least damaging to the efficiency of the system. The solution is to consider multipart tariffs and differentiating pricing according to willingness to pay, but with marginal social cost as a starting point, and then looking for the optimal departures from that base. But, unless multipart tariffs can be found, which exclude no users and face all users with a marginal price equal to marginal social cost, there will be an efficiency loss due to these measures (except in circumstances where there is a budgetary problem and there is no more efficient way to raise the necessary revenue from general taxation). That efficiency loss will be minimised by applying the budget constraint at the most aggregate level possible, so budgetary problems will not lead to a case for requiring each mode to cover its total cost from revenue.

(e) The institutional issues concern ways of providing appropriate incentives for the efficient provision and development of transport infrastructure. There are two crucial institutional issues surrounding marginal social cost pricing. The first is the general one of cost coverage. Rothengatter (2003) asserts a general rule that charging users the full cost of the infrastructure rather than subsidising it will lead to technical and allocative efficiency in the level of infrastructure provision. In the presence of economies and diseconomies of scale, as discussed above, there will be a cost in terms of efficiency if full cost recovery suppresses or expands demand compared with the level that would arise with marginal social cost pricing. This cost could be more than offset if it were the case that the full cost recovery rule led to an increase in technical efficiency of the infrastructure provider. But it is unclear why this should be the case. Most transport infrastructure is a natural monopoly, which may use its monopoly power in pricing to break even without being technically efficient; on the other hand, given the presence of economies of scale, there may be economically efficient infrastructure investments that are unable to break even with any practicable tariff.

The second concerns the incentives to infrastructure providers given by short-run marginal social cost pricing. Under short-run marginal social cost pricing, prices will be high where capacity is inadequate, giving the infrastructure manager an incentive to withhold necessary investments. Of course, if it could be assumed that infrastructure capacity decisions were taken by governments, solely on the basis of social cost–benefit analysis, then this would not be a problem. But most economists recognise that governments are not so entirely benevolent in their motives. If it is considered that efficiency requires infrastructure managers to be constituted as commercial organisations exploiting private capital to fund their investment, then the problem of reconciling incentives to infrastructure managers with efficient pricing becomes acute. Again, long-run marginal social cost pricing has attractions in this case, but given that transport infrastructure managers always have considerable monopoly power, a regulator will still be needed to ensure that investment levels are efficient and that charges are related to the costs actually incurred. Hence, to the extent that infrastructure is a natural monopoly, it is difficult to see any approach to optimal provision and pricing that does not have some degree of government intervention; whether that intervention should best take the form of regulation of private monopolies or direct provision and control remains a major source of controversy.

(f) It is long established that the case for marginal cost pricing assumes the existence of marginal social cost pricing for related goods elsewhere in the economy. For instance, if one good is charged a price below marginal social cost, there is a case for charging substitute goods below marginal social cost as well to reduce distortion of choices between them. The practical implication is that, where there are divergences between price and marginal social cost in related markets, these lead to cases for divergences in the market in question. The most common application of this argument is in terms of competition between modes. But there are other concerns. For instance, if the costs of urban sprawl are not adequately reflected in property prices and local taxation, then it may be desirable that transport prices are configured in part to offset this distortion and to discourage sprawl. The implication is that charges for entering town centres should be below marginal social cost. Similarly, if labour taxes discourage labour supply that may lead to a case for subsidising the cost of commuting.

Thus, in essence, most advocates of marginal social cost pricing would accept these criticisms. What they would say is that 'pure' marginal social cost pricing has to be modified to take all these issues into account. This is the subject of second best pricing, on which a wide literature exists (Verhoef, 2001). In essence, the simple marginal social cost pricing rule has to be modified to reflect divergences between price and marginal social cost elsewhere in the economy, weighted by the degree of interaction between those sectors and the service for which the price is being set. The resulting pricing rules may be very complex, but it is likely to be better to make the best possible estimate of the necessary data than to ignore the issue.

(g) Regarding the final criticism of marginal social cost pricing, it is certainly true that implementation of sophisticated pricing systems remains far from costless, whilst the ability of users to respond to the prices may be limited by their ability to understand and predict what they will have to pay. Thus there is an optimal degree of complexity of any price structure arrived at by comparing the benefits of greater differentiation in terms of their influence on the volume and location of traffic with the costs. There are also issues concerning the speed of implementation, which may differ between the modes both because the technical requirements differ and for political reasons. In turn, these factors lead to second best considerations; differing degrees and speed of implementation between the modes could itself be distorting and needs to be taken into account in policy determination. Thus, implementation of marginal social cost pricing is far from straightforward, but there seems to be good evidence that simple systems of road pricing in heavily congested cities or on congested inter-urban links can bring worthwhile benefits (Nash, Niskanen, & Verhoef, 2003), whilst more complex systems are becoming steadily easier to implement as technology advances.

To sum up, then, Rothengatter does well to remind us of the whole range of factors that mean that pure marginal social cost pricing is not a desirable or sensible aim to follow. The set of issues he raises is very important, and no doubt the solution to them will include multipart tariffs in appropriate circumstances. But this does not mean that a totally different theoretical approach to pricing policy needs to be adopted, or that full cost recovery as a principle is a good starting point. It is possible to measure marginal social cost and to move towards it as the basis of transport pricing although difficulties and uncertainties remain. Considerations such as budget constraints, equity, institutional issues, simplicity and price distortions elsewhere in the economy lead to a need to depart from pure marginal social cost pricing but do not change the position that the measurement of marginal social cost is the correct starting point in the development any efficient pricing policy. For this reason, recent projects for the Commission that have sought to address these issues have tended to use the phrase 'marginal social cost-based pricing' rather than 'marginal social cost pricing' to summarise the philosophy they adopt (Verhoef, 2001).

4. DEVELOPMENTS IN EC TRANSPORT PRICING POLICY

One arena in which theory has come close to being transferred from the text book and implemented in transport policy has been at the level of the European Commission. Having set out, in its Green Paper on 'fair and efficient pricing in transport' (CEC, 1995a), the basic principles it proposed to follow, it subsequently set out its strategy for pursuing those principles in a White Paper entitled 'Fair Payment for Infrastructure Use'. The core features of the White Paper focused on the need to relate charges more closely to the underlying marginal social costs associated with infrastructure use, extending these costs to include external costs, and with the need to depart from prices that are purely based on the direct costs of infrastructure use when cost coverage requirements need to be met. The need to ensure transparency, and to facilitate fair competition between modes, within modes and across user types was emphasised. Furthermore, the contribution of transport services to the enhancement of industrial efficiency and European competitiveness was recognised.

In order to give transport users and providers time to adjust, the White Paper proposed a phased approach to the implementation of this framework. The first of three phases, identified as running from 1998 to 2000, aimed to ensure that a 'broadly compatible structure is in place in the main modes of transport' (CEC, 1998). Air and rail were to be the particular focus of this first phase; charges incorporating external costs, on the basis of an agreed Community framework, were to be allowed, but total charging levels were to be capped by average infrastructure costs. The second phase, identified as running from 2001 to 2004, aimed to oversee further harmonisation. The White Paper proposed that this phase would particularly focus on rail and heavy goods vehicles, for which a kilometer-based charging system differentiated on the basis of vehicle and geographical characteristics was proposed, and on ports, where a charging framework would be introduced. From here on in, it was proposed that charges should be capped at marginal social cost. The third and final phase, identified as running from 2004 onwards, focused on revisiting the overall charging framework, with a view to updating it in light of experience.

The principle of subsidiarity, which recognises that the location-specific nature of many transport externalities means that policy action is often better pursued at the national or local level, rather than the European level, was affirmed by the Green Paper on fair and efficient pricing. This has meant that European policy development has focused much less on urban transport than on inter-urban transport. However, the Green Paper did highlight the possible need for European involvement in local issues where they affected the efficient workings of the internal market. The White Paper (CEC, 1998) went on to commit to encouraging member states to develop urban road charging systems and to reviewing any Community legislation that may harm implementation. In furtherance of its plan to encourage member states to develop urban road charging, the Commission has supported and facilitated a number of cross-national networks of interested cities (e.g. EUROPRICE and PROGRESS). In addition, the 'Citizen's Network' initiative (CEC, 1995b) sought to promote the concepts of affordability and quality in local and regional public transport and included pricing as one of a number of practical measures for the promotion of sustainable transport.

Regarding railways, Directive 91/440 sought to separate accounting for railway infrastructure and operations in order to make the basis for railway infrastructure charging transparent, whilst opening access for specific types of international services. The more recent directive on rail infrastructure charging (2001/14) requires marginal social cost to be used as the basis of charging, whilst permitting supplementary charges where necessary for cost-recovery purposes (European Parliament and Council of the European Union, 2001).

The review of Transport Policy (CEC, 2001) reaffirmed the commitment to more efficient pricing of transport in order to internalise externalities, and proposed a framework directive on pricing to set out the principles to be followed in all modes of transport. This document also identified an important link between pricing and financing, and proposed to permit funds raised from some sectors of the industry to be used for worthwhile projects in other sectors where the result is to reduce social costs. Thus, for instance, charges raised for environmental costs of road transport may be used for new rail infrastructure, as in the explicit linking of new hgv charges in Switzerland to the funding of the new rail tunnels under the Alps. This will require replacement of the Eurovignette Directive by one permitting the charging of all social costs and the use of the revenue in this way.

The 'Eurovignette' directive 1999/62/EC aimed to limit competitive problems within the road freight sector caused by the existence of very different methods and levels of charging for infrastructure use in different countries. For example, vehicles licensed in a country with low annual licence duty plus supplementary tolls may have an unfair competitive advantage when competing with a vehicle licensed in a country with high licence duty and no supplementary tolls. The Eurovignette was intended to set a limit for the maximum infrastructure access charges payable as a general supplementary licence for heavy goods vehicles, on the basis of average infrastructure costs, with non-discrimination between goods vehicle operators of different nationalities.

A proposal to amend Directive 1999/62/EC on the charging of heavy goods vehicles for the use of certain infrastructure was put forward in 2003. The overall average charge had to be equal to infrastructure and uncovered accident costs (i.e. external cost of accidents minus insurance premiums), where infrastructure costs must be allocated to vehicle types on the basis of stipulated equivalence factors and may include the annualised cost of investment going back 15 years only.

The directive proposed that the charging system should be more differentiated and tolls could vary according to a number of factors such as the distance travelled, the location, infrastructure type and speed, the vehicle type (axle weight, engine type, energy source, emission standards), the time of day and specific routes (CEC, 2003). The toll will be applied to HGVs weighing over 3.5 tonnes, replacing the previous 12 or more tonnes. The earlier Directive will change by being applied to the TEN network and to other roads to which traffic might divert, but permitting application of pricing to other roads as well (the previous directive only covered motorways).

Changes to the use of revenue from the tolls according to the proposed directive, include ensuring strict enforcement that revenue from the tolls is used for expenditure on roads, other transport networks, transport substitutes or the transport sector as a whole, but not general state expenditure such as spending on health or education. Each state must create an independent transport infrastructure supervisory authority to guarantee that charges are being set and revenue is being used in the required way. In exceptional circumstances a surcharge of up to 25% will be permitted to fund alternative rail infrastructure.

Despite the flurry of activity at the level of the Commission, progress on implementation of the policy has been slow. The proposed framework directive and methodology paper have not been forthcoming. Even in the rail sector, where infrastructure charging is generally a new concept resulting from the separation of infrastructure from operations, a variety of approaches to charging has been taken in the different member states. UIC has argued strongly that this is a major barrier to the success of rail in international transport (Gustafsson & Knibbe, 2000). At the time of writing, no agreement was reached on the proposals regarding heavy goods vehicle charging, which itself continues to limit the ability to price at marginal social cost, whilst permitting more efficient differentiation than is currently the case. On air and water transport there is a lack of clear progress. The result is that research has focussed increasingly on implementation issues and understanding the barriers to progress and how they may be overcome. However, the number of countries now implementing or considering a kilometer-based charging system for heavy goods vehicles may indicate that the pace of implementation is accelerating.

5. RESULTING RESEARCH NEEDS

The 1995 Green Paper on Fair and Efficient Pricing triggered a major effort in research on transport pricing, much of it funded under EC Framework programmes. Numerous research needs were identified at that time and as progress has been made on those, new research needs have emerged.

The first set of projects (including particularly PETS, QUITS and TRENEN) concentrated very much on developing pricing principles, on measuring the various elements of marginal social cost and on case studies to model the impact of its implementation. These were brought together in a Concerted Action that sought to build up consensus on these issues (CAPRI).

Since then, there has been a much greater emphasis on implementation issues rather than basic principles. AFFORD and MC-ICAM, for instance, identified the need for policy packages, both to ensure acceptability and to make the most efficient use of revenue. MC-ICAM also adopted the idea of a phased approach, as set out in the 1998 White Paper, and studied the reasons why such a phasing might be necessary, in the light of the constraints provided by barriers to implementation, and the way in which these might change over time. As a result, it came up with the most efficient and equitable feasible implementation paths. REVENUE is specifically examining the use of revenue from transport pricing, whilst PROGRESS, CUPID and DESIRE concern themselves with the practical issues of implementing road pricing in urban and inter-urban areas respectively. Again, the work of these projects was synthesised and presented to policymakers through a thematic network named IMPRINT-EUROPE.

The optimal degree of complexity of transport charges is a new and largely unresearched issue that is crucial for implementation. Previous work has identified that charges would need to be complex to reflect marginal social cost accurately, and that simple pricing schemes are more acceptable than complex ones. But two further issues need to be brought into the picture. One is the relative cost of implementing schemes of varying complexity and the way in which this is varying over time. The second is the degree to which people correctly perceive and act on pricing schemes of varying complexity.

Thus a wide range of issues are on the transport pricing agenda. Fundamental to all, however, is the need for credible estimates of marginal social cost. Without such estimates much of the remaining research is of limited value, i.e. for example, if we do not know what marginal social cost is, then how would we know what a price based on marginal cost would be, and how would we know what impact such a price might have?

It is the UNITE project that provides the research forming most of the content of this book. UNITE sought to fulfil the requirements of transport decision-makers for information for pricing decisions. The information required was seen as comprising two types – detailed case studies of marginal social cost for specific circumstances, and broad estimates of overall social costs and revenues for national or regional transport systems as a whole. The third major element of the UNITE project was to study ways of using these two sorts of information together to derive appropriate policy conclusions regarding pricing systems that were both efficient and equitable.

6. OUTLINE OF THE BOOK

Following this introduction, Chapter 2 provides a more detailed account of alternative pricing doctrines in the transport sector and of their influence on pricing policy. We then review the state of the art in the measurement of the marginal social cost of transport, looking in turn at infrastructure, operating costs, user costs (both of congestion and of charges in frequency of scheduled transport services) accidents and environmental costs. These chapters (3–7)draw heavily not only on the work of UNITE, but also on other projects where appropriate. In this book we cover both urban and inter-urban transport, and the road, rail and air modes of transport. We do not in general cover waterborne transport; this was not tackled in depth in UNITE and relatively little appears to be available in terms of other studies of marginal social cost. This is therefore a priority for future research.

We then turn to the use of the estimates. Chapter 8 examines the ever-growing evidence on the impact of marginal cost pricing, using estimates of marginal social cost in forecasting models. Chapter 9 reports on a closely related project called RECORD-IT, which examined the particular issue of inter-modal freight, providing a methodology for estimating the marginal social costs of the various modes and combinations of modes involved to see to what extent taking account of externalities improved the case for inter-modal systems. Chapter 10 addresses the difficult but all important issue of how to generalise from the results of case studies to obtain estimates of marginal social costs for all circumstances. Finally, Chapter 11 presents our conclusions.

Those wishing to look into the results of the projects discussed in this book in more detail are advised to start with the website www.imprint–eu.org. The seminar papers found on this website review most of the projects discussed in this book, and there are links to other websites including that of UNITE.

REFERENCES

Commission of the European Communities. (CEC). (1995a). *Towards fair and efficient pricing in Transport*. Brussels.

Commission of the European Communities. (CEC). (1995b). *The citizen's network. Fulfilling the potential of public passenger transport in Europe*. Brussels.

Commission of the European Communities. (CEC). (1998). *Fair payment for infrastructure use: A phased approach to a common transport infrastructure charging framework in the EU*. Brussels.

Commission of the European Communities. (CEC). (2001). White paper: *European transport policy for 2010: time to decide*. Brussels.

Commission of the European Communities. (CEC). (2003). *Proposal for a directive of the European parliament and of the council amending Directive 1999/62/EC on the charging of heavy goods vehicles for the use of certain infrastructures*. Brussels.

European Parliament and Council of the European Union. (2001). *Directive 2001/14/EC on the allocation of railway infrastructure capacity and levying of charges*. Brussels.

Feldstein, M. (1972). Distributional equity and the optimal structure of public prices. *Am. Econ. Rev.*, *62*, 32–36.

Gustafsson, G., & Knibbe, A. (2000). Infrastructure Charges in Europe – a missed chance to increase competitiveness. Discussion Paper 245, Helsinki Workshop on Infrastructure Charging on Railways, VATT, Helsinki.

Lindberg, G. (2002). Recent progress in the measurement of external costs and implications for transport pricing reforms. Paper presented at the second Imprint-Europe seminar, Brussels.

Mayeres, I., & Proost, S. (2003). Reforming transport prices: An economic perspective on equity, efficiency and acceptability. Paper presented at the fourth Imprint-Europe seminar, Leuven.

Mohring, H. (1972). Optimisation and scale economies in urban bus transportation. *Am. Econ. Rev.* Papers and Proceedings (pp. 591–604).

Nash, C. (2003). Marginal cost and other pricing principles for user charging in transport: A comment. *Transport Policy*, *10*, 345–348.

Nash, C., et al. (2003). *UNITE (Unification of accounts and marginal costs for transport efficiency) final report for publication*. Funded by 5th Framework RTD programme.

Nash, C., Niskanen, E., & Verhoef, E. (2003). Policy conclusions from MC-ICAM. Paper presented at the 4th Imprint-Europe seminar, Leuven.

Rothengatter, W. (2003). How good is first best? Marginal cost and other pricing principles for user charging in transport. *Transport Policy*, *10*, 121–130.

Roy, R. (2002). The fiscal impact of marginal cost pricing: The spectre of deficits or an embarrassment of riches? Paper presented at the second Imprint-Europe seminar, Brussels.

Verhoef, E. (2001). Marginal cost based pricing in transport: Key implementation issues from the economic perspective. Paper presented at the first Imprint-Europe seminar, Brussels.

ALTERNATIVE PRICING DOCTRINES

Emile Quinet

1. INTRODUCTION

Transport infrastructure pricing has long been the subject of discussions among lobbyists and analysts in Europe. Most people agree on the necessity to reach an agreement to harmonise infrastructure pricing at the level of the European Union but very few agreements have been reached during the past 50 years, despite the concerted efforts of the European Commission over the past decade. Where agreements have been reached, they have taken a long time and a great deal of effort to achieve and, in the end, involve considerable compromise and imply little actual change. It seems that those engaged in the debate are repeating their own positions without managing or even trying to understand the alternative positions, and it is contended here that this lack of understanding is probably a major reason for the curious situation we find ourselves in.

It is therefore interesting and useful to try and compare the doctrines at stake throughout the continent, a task to which this chapter is devoted: what doctrines are supported by the various actors (political bodies at the national or European levels, administrations, transport professionals, academics), what factors underpin these doctrines, what judgement can be borne on them from the point of view of economic analysis, and are these doctrines actually so different and incompatible? By doctrine, we mean a set of ideas, which is generally supported by the country or by the group of people under consideration, based on some considerations of a theoretical

Measuring the Marginal Social Cost of Transport
Research in Transportation Economics, Volume 14, 19–47
Copyright © 2005 by Elsevier Ltd.
All rights of reproduction in any form reserved
ISSN: 0739-8859/doi:10.1016/S0739-8859(05)14002-5

nature or on empirical evidence, or more often on both, and which becomes a kind of conventional wisdom.

The rest of the chapter is organised as follows: Section 2 recalls the doctrine of the European Commission, which is based on the Short-Run Marginal Cost (SRMC) principle, and lists the arguments based mainly on efficiency concerns, in favour of this position. This doctrine is a benchmark as it is proposed and supported by the Commission, and has given way to many debates; thereafter, any position in the field of infrastructure pricing can and even should be analysed by reference to this doctrine.

Section 3 gives an overview of alternative pricing doctrines in a selection of European countries. This overview shows that the current views about charging are based mainly on principles such as Average Cost (AC) or Development Cost (DC) or Long-Run Marginal Social Cost (LRMC), and differ from the doctrine of SRMC. The motivations for the deviations from the theoretical prescriptions are then analysed. It is shown that most of the concerns that lead to alternative solutions are valid.

However, it is argued in Section 4 that economic theory is able to deal with these concerns and that therefore, alternative solutions are not called for. Furthermore, the situations where the most appropriate charges differ from SRMC are rather infrequent.

These results are synthesised in the conclusion (Section 5). It is argued that the distances between these various doctrines are probably not so large compared to their distances from the present situation, and that any kind of movement towards the recommendations of economic analysis would be a great improvement.

2. THE BENCHMARK: THE MARGINAL SOCIAL COST PRICING PRINCIPLE

The European Commission's proposals for infrastructure charging reforms have been expressed in many documents, but most extensively in the now well-known Green Paper of 1995 and White Paper of 1998 and, more recently, the 2001 White Paper. The proposals contained in these documents are based on the principle of SRMC pricing, which implies that each user of transport infrastructure should pay – at or close to the point of use – the full marginal social cost imposed by that use.

The well-known traditional justification of marginal social cost pricing is that it is "allocatively efficient" in the sense of optimising the allocation of resources and thus maximising the welfare of society as a whole. The argument, founded on well-established principles of economic science, is summarised in the White Paper:

> 'Where charges are too low, excessive demand is likely, generating higher costs than benefits, and individual operators have less incentive to reduce the costs that they impose on society. Where charges are too high, some users who would be able to pay the costs they impose would be discouraged from using the infrastructure, thereby reducing its social benefit'.

This result can be rigorously demonstrated through several theoretical presentations (see, for instance, Laffont, 1984), these presentations enlighten the conditions of validity of this result: as a whole, it is valid in situations which are called "first best" by economists, i.e. when markets are competitive and when there is no external effect nor fixed costs. When these conditions are not fulfilled, economic theory provides indications on the corrections to apply to the pure and strict marginal social cost pricing principle.

These results are valid for any kind of good, but the adaptation to the case of transport must take into account the fact that the cost of transport comprises not only the cost for the operator or infrastructure manager, but also the cost incurred to the other members of society; this means that, over and above the marginal operating cost of each trip, each user – in practice: each transport unit (truck, car, bus, train, etc.) – should pay for:

- the marginal cost of infrastructure damages (the increase of infrastructure cost due to an extra traffic unit)
- the marginal external cost of congestion and scarcity (the extra cost due to the introduction of an extra traffic unit, either as an increase in cost or as a reduction of benefit for other users
- the marginal external cost of pollution, the increase in total pollution costs due to an extra traffic unit
- the marginal external cost of accidents, which is both marginal (the increase of cost for an extra user) and external (this increase is related to costs not borne by the user).

A first important point to notice is that it is a short-run cost:

> 'Marginal costs are those variable costs that reflect the cost of an additional vehicle or transport unit using the infrastructure. Strictly speaking, they can vary every minute, with different transport users, at different times, in different conditions and in different places' (EC White Paper, CEC, 1998).

Another important point is that, in the previous considerations, the cost was the cost of the extra-unit, i.e. the marginal cost; it implies that the charges should not include the fixed costs of infrastructure provision, nor any other taxes over and above the applicable rate of VAT. This point may lead to deficits for the infrastructure manager, the revenue from the charges being lower than the expenses. In order to cope with this point, generally recognized as a drawback for the management of infrastructure, the White Paper acknowledged the need for departing from SRMC – through two part tariffs or similar devices – aiming at cost recovery strategies for infrastructure terminals and some new infrastructure.

Following the exposition of these principles and how they might relate to transport infrastructure, many studies have been undertaken aimed at allowing for workable and common methodologies for their implementation, establishing what concrete pricing instruments are most efficient in different contexts and what the appropriate price levels are. Furthermore, in order to have public and political acceptance for the suggested pricing principles, more knowledge on the impacts – costs and benefits – of pricing and the relationship of pricing to financing was also required. Research activities under the 4th and 5th Framework Programmes looked into most of the above-mentioned questions and issues, through projects such as CAPRI, DESIRE, AFFORD, MC-ICAM, IMPRINT and, more recently, UNITE.

Though inspired by economic analysis, this principle of marginal social cost pricing is not accepted by all actors of the transport sector. Either groups of people, such as transport industry or countries, support alternative doctrines. Why are there such alternative doctrines, who supports them, what is their content? The next section intends to take an overview on these questions.

3. AN OVERVIEW OF PRICING DOCTRINES IN A SELECTION OF EUROPEAN COUNTRIES

In order to take a picture of the variety of doctrines, a review was undertaken of the opinions and doctrines prevailing among actors in the industry, namely political decision-makers, administrators, transport operators and academics. The present section first provides information on the review, second, outlines the results, and finally analyses the justifications of the various doctrines, mainly stressing the differences in the political doctrines expressed in different countries.

3.1. A Review of Doctrines Employed in Teaching, Policy and Practice

As there was, to our knowledge, no existing review of the doctrines, a small survey was conducted among scholars in a selection of European countries: Austria, France, Germany, Ireland, Spain, Switzerland and the United Kingdom.

Three questions were asked, in an open questionnaire:

- What are the differences between the picture given by the theoretical review and the current teaching at universities about transport?
- What are the current doctrines expressed by the political authorities (Government, Parliament, etc.) on the subject of transport infrastructure pricing?
- What is the real situation of present infrastructure pricing?

The survey was performed in a very simple way: the questions were addressed to a scholar of each of the countries (mainly those involved in the UNITE programme, in the framework of which the study was undertaken). The answers were synthesised and sent back to the respondents[1] for control. A summary of the answers is presented in the appendix.

It is possible to synthesise these answers and to classify the various positions which are expressed; the classification is made according to different keys for each of the three topics.

3.2. Differences about Teaching

Table 1 presents the results relating to teaching as a whole. It can be seen that differences in teaching lie between basic courses in economics or engineering and advanced economics courses; it appears that standard economic theory – i.e. the SRMC principle with its validity limits – is taught only in the more advanced economic programmes in universities (in many countries, it is considered as mostly theoretical and difficult to apply); in other courses, such as engineers courses or MBA in universities, less sophisticated methods are taught, based on cost allocation procedures (this point probably comes from the fact that, in the beginning, transport economics was developed by engineers).

It follows that the SRMC principle is recognised as a sound principle by economic teaching, most courses indicate also its theoretical validity limits; high level programmes develop methods to overcome these limits; but engineering teaching sticks to the "traditional" cost allocation procedures.

Table 1. Differences on Infrastructure Charges in Teaching.

Doctrines → Type of Teaching ↓	Cost Allocation Procedures	SRMC Principle	More Elaborated Doctrines
Engineering economics	Yes	Sometimes	No
Basic economics (e.g. MBA)	Hints	Yes	Hints
Advanced economics	No	Presented as the basis	Yes

3.3. Differences about Political Doctrines

Differences between doctrines currently expressed by political authorities appear between different countries:

- France and the United Kingdom are the most in favour of marginal cost pricing, mainly in the shape of LRMC: the French standpoint is mainly due to the fear of strategic behaviour by operators (SRMC is deemed to induce low charges leading to increased traffic and calls for more investment) and to equity considerations (with LRMC, the users pay for the whole set of expenses, not only for those which are variable); in the United Kingdom, although the traditional approach to public utility pricing has been LRMC, an interest in congestion pricing is developing, for instance, through the urban road pricing schemes, as in London: this is clearly SRMC in nature.
- In other countries, doctrines about charging stem mainly from financial considerations, first, in relation to the overall public budget and, second, in relation to the transport sector budget; this position is expressed in Austria (environmental issues are dealt with through regulations; charges are aimed at raising funds for general and transport budgets) and in Ireland (where no momentum towards road pricing was found).
- In a third category of countries, close to the second, doctrines are based on financial considerations, but primarily in relation to the transport sector budget and, second, in relation to the overall public budget: in Germany, the charging principle is based on average cost, with an interest in the Polluter Pays Principle (PPP) in relation to the environment (acknowledging that environment is mainly dealt with through regulations); the standpoint of Switzerland is similar, with the fact that efficiency considerations, which are at the basis of SMC pricing, do not seem to be an

Table 2. Differences in Political Doctrines on Infrastructure Charging.

Doctrines → Countries ↓	Public Finance Considerations	Cost Allocation Procedures Inside the Transport Sector	Concerns about Environmental PPP	SRMC	LRMC	More Elaborate Principles
Austria	X		X			
Germany, Switzerland		X	X			
The U.K.			X	X	X	
France	X				X	
Nordic Countries			X	X		X

important issue; in Spain there is an emerging interest in infrastructure charging, due to financial concerns and to the idea that users should pay more than they presently do for the use of infrastructure.

- Finally, though they had not been included in the formal review, it must be noted that Nordic countries are in favour of environmental protection through the use of environmental taxes and have espoused the marginal social cost principle, though many of their initiatives are inspired by public finance considerations.

These results are summarised in Table 2. This is, however, very tentative, as in most countries, there is no clear expression of official political will.

3.4. Differences between Infrastructure Pricing Practice and Principles

The main differences between infrastructure pricing in practice are in relation to modes. In all countries, the types of charges, which exist for one mode or another, are roughly the same, but it is the differences between different modes that are very important, so the presentation here is done by mode, though in each case the (generally small) differences between different countries are also indicated.

- *Road:* the most used means of road charging are parking fees and annual vehicles taxes, often at a local level, and fuel taxes (these are partly earmarked in Germany and Switzerland). There is a general tendency towards mileage-related charges, proposed in Germany and Great Britain (first for heavy goods vehicles, then possibly to be extended to other types of vehicles) and already implemented in Austria and Switzerland. At the time of the survey, there were already vignettes for access to the motorway network in Austria and Germany – where the vignettes are limited to heavy goods vehicles – and in Switzerland – where vignettes existed for any kind of vehicle. The Austrian and Swiss systems have since been replaced by mileage-related charges for heavy goods vehicles, while a similar system in Germany has been delayed by technical problems. Vignettes and mileage-related charges are more or less linked to pollution and can be considered as an application of the PPP. In France and to a lesser extent Spain, the charging system relies heavily on tolled motorways, the tolls being calculated for financial, rather than efficiency, purposes. There are tolled bridges and tunnels in every country, but they are especially important in Austria and Switzerland for Alpine transit. Let us note also two emerging facts: first, the proposal of the Commission for a revised Eurovignette which would lead to a mileage-related charging system; second, the electronic GPS charging system under implementation in Germany.

- *Rail:* The charging systems for rail are recent; they date back only to the introduction of competition within the last 10 years and there are large discrepancies between countries, concerning both the principles and the results. Cost recovery is high in Austria, United Kingdom and Germany, and in the last two countries tariffs were two-part tariffs with a high fixed part, although in Germany this has since been abandoned to take into account the protestations of new entrants. More recently, cost recovery has been reduced in the United Kingdom, requiring public subsidies. In France, the charging system does not cover more than half the expenses and is based on parameters and relations, which are similar to those which prevail in SRMC. An infrastructure charging system is to be implemented in Spain and Switzerland, and should cover at least the SRMC. Beyond the differences in doctrines, there is a large heterogeneity in infrastructure charges in Europe (see, for instance, Peter, 2004).
- *Inland waterways:* The charging system is aimed to cover the total costs in Austria, Germany and Switzerland; charges are very low in France and probably do not cover the SRMC.
- In Air Transport, as well as in Sea Transport, the general principle is a total cost recovery, coherent with the fact that infrastructure operators of these modes are (often public) firms having to break-even. But the application of this principle can lead to many arguments as, in the majority of countries, these firms benefit from public subsidies, the fairness of which is not always assured.

Concrete charging systems are the result at a point of time of a stratification of various past decisions, taken by different non-coordinated decision-makers. For instance, local authorities fix parking fees and national authorities determine fuel taxes. These decision-makers are moving over time and are subject to electoral agendas and pressure from lobbies.

3.5. Criticisms of SRMC Pricing

In this general framework, from the previous survey and also from the author's experience and discussions with experts, it appears that decision-makers find several drawbacks to SRMC; let us list them according to the actors:

- *Equity*. Political decision-makers are not concerned much by efficiency as they derive no benefit from a gain in efficiency, which affects the voters only to a limited extent. They are, however, very interested in equity,[2] as are the voters. It appears that SRMC has many drawbacks on this ground; for instance, congestion costs induce a very high charge during peak-hours, when the majority of trips are home-to-work, and hit low income groups.
- *Acceptability* is another important political concern. This notion is close, but not exactly equivalent to equity. It is related to the bargaining power of social groups, which struggle against a decision which would hurt them (unfair measures can be accepted if the social groups which are disadvantaged have no bargaining power). It turns out that, in general, the consequences of SMC pricing have a low acceptability level, as is shown by the unsuccessful proposals of road pricing in several European countries.
- *Deficits*. It is often thought that SRMC leads to deficits, due to fixed costs and increasing returns to scale. It may be deemed as being unfair if, where there are deficits these are funded by tax-payers whilst the people who benefit – the users – are only paying a low charge. Furthermore the deficits have to be funded through an increase of taxes, a point which is seen not only as politically harmful, but also as leading to inefficiencies: decision-makers know, as well as economists, that a body which is allowed subsidies is induced to inefficient behaviour and spends relatively more time and effort in rent-seeking than in cost abating.
- *Uncertainty and information issues*. On top of that, all actors consider that SRMC calculations are difficult and uncertain, and that a lot of expertise

and audit is necessary to check their validity. In the real world, decisions are the result of bargains between several pressure groups and stakeholders, who use the uncertainty of the calculations to try to reach their private goals.

These points are stressed by many actors in the transport sector. Those who are dealing with transport management on the professional and administration sides are historically engineers and they developed a corpus of ideas inspired by logic, wise spirit and a sense of equity; they are well acquainted with accounting practice in normal industrial activities and tend to transpose them in transport, and use them for infrastructure pricing. These actors think that SRMC leads to deficits, a point that makes it useless for pricing as it would be at odds with private practices.

3.6. Alternative Concepts

These reasons explain why SRMC is often rejected to the benefit of concepts such as LRMC, DC, AC or *Full Cost*. Though they are generally not precisely described, they cover the idea that the users should pay for the expenses they cause to society.

The LRMC is the derivative of the infrastructure long-run cost function (in the long-run cost function, it is assumed that the investment level is optimised for the traffic level under consideration) with respect to the traffic under the assumptions that infrastructure capital is optimally adjusted. In the case of transport, infrastructure does not vary continuously. Apart from this problem of indivisibility when infrastructure is optimally adjusted, the pure LRMC is equal to the SRMC; elsewhere it is not defined.

A practical implementation of LRMC which overcomes the problem of indivisibility is the DC. It is the ratio between the discounted sum of the future investments and the discounted sum of the traffic increases that make them necessary, both being taken over a long time period. On the one hand, DC has the virtue to include investment in the charges and to make the users sensitive to the investment expenses; furthermore it smoothes the SRMC, and makes the charges more even across the years and the specific situations over time and space. On the other, as the charges are smoothed, the incentives to the users, for peak periods for instance, are less powerful; besides the virtues of the development cost have not been explored by economic theory.

Other advocated concepts are the concepts of AC or Full Cost. A very wide panoply of calculation procedures have been developed around these concepts. Several options have been discussed in relation to them. The first ones relate to the numerator side: which expenses have to be distributed across the various categories of traffic: actual transport expenses, the actualised historical construction expenses, or the expenses that would be incurred if it were necessary now to build and operate a modern piece of infrastructure? Other options relate to the denominator side: how to distribute the cost between the different categories of traffic? An overview of past and current practices learns that two types of solution are often used. Firstly, there are accounting-type solutions, based on equivalence ratio between traffic categories for the various kinds of cost categories. For instance, pavement thickness is allocated according to the damage caused by axle load (4th power of the axle load according to the AASHO tests, based on Highway Research Board, 1962). Secondly, there are solutions based more explicitly on co-operative game theory. The idea is to find an allocation, which is in the core of the co-operative game of sharing the cost of the infrastructure, so that no traffic would have interest to leave the coalition to set up its own infrastructure. Procedures such as the weighted average of all possible incremental costs are an example of solutions advocated by the game-theoretic approach (Curien & Gensollen, 1992; Castano-Pardo & Garcia-Diaz, 1995). These procedures were the dominant doctrine about 20 to 30 years ago in every country. They still have supporters. For instance, in the U.S.A., cost allocation studies based on these principles are regularly published (see Link, Stewart, Maibach, Sansom, & Nellthorp, 2000).

The ideas that support these concepts are manifold and are related to the concerns of decision-makers and to the drawbacks they see in the SRMC principle.

A first reason for advocating concepts such as DC, AC or Full Cost is related to the difficulties of SRMC calculations and to the possible manipulations by pressure groups and lobbies. In comparison, concepts such as AC or DC (this excludes the external costs, especially congestion costs) seem more simple and less uncertain, and therefore less manipulable.

Other considerations are based on efficiency incentives for the operator: SRMC pricing does not screen unprofitable services with high fixed costs which are not incorporated in the charge, and the operator can use this fact and the asymmetry of information to manipulate the cost, lowering the marginal cost in order to increase the patronage and gain more subsidies from the public authorities.

Other concerns are related to problems of equity. First, SRMC pricing is often seen as not providing enough funds to finance the total expenses. Second, SRMC pricing is not seen as an equitable solution, as it leads to high fares being charged on captive users, for instance the commuters who are relatively low-income users, and cannot cope with public service obligations such as low fares for redistribution objectives.

AC seems to avoid problems of complexity and uncertainty in calculations, lack of finance and manipulations of fixed costs and subsidies: if fixed costs are too high, the AC will be high too. Eventually, because of the increase of the charge, the demand will disappear, causing the closure of services whose fixed cost is too high. It solves also some equity problems in the sense that it ensures that transport costs are paid by the users and not by the taxpayers. The problem is that average cost is arbitrary, as there is no non-arbitrary way of allocating the common costs (the procedures that have been already quoted have no logical justification), except if the allocation of common costs is made according to the Ramsey rule, which is based on SRMC (the Ramsey rule is a solution of the problem to optimise collective welfare, the infrastructure manager being subject to a balanced budget constraint; the result is that, under some hypotheses, the relative discrepancy between marginal cost and optimal charges is inversely proportional to the demand elasticity).

The problem of manipulation of SRMC for rent-seeking behaviours is real, and AC pricing is a way to fight against it. But it is clear from the above presentation that AC also has a lot of uncertainties, especially for the breakdown of total expenses between the categories of traffic.

The LRMC also seems to be easier to calculate than SRMC as it does not take into account congestion cost; furthermore, it corresponds more or less to the idea that SRMC leaves aside investment costs, which is necessary to take into account. When saying this, people are not fully aware that LRMC equals SRMC in the optimal situation and does not exist in other situations, and that it does not easily take into account changes in quality of service. Another point is that it is not easy to determine the optimal investment, and that the result depends on the investment level.

DC relies on the same reasoning. It looks smart and it is attractive because it seems to combine several nice features (the word marginal is avoided, the reference to investment, the relation to the expenses). It avoids the objection not to be defined when the situation is not optimal, but it has no real justification.

In conclusion, it appears that many of the concerns of the decision-makers about SRMC are quite valid. But in order to deal with these concerns, they suggest alternative solutions, which do not really solve the problems they intend to solve and which have several drawbacks themselves.

4. ARE THESE ALTERNATIVE CONCEPTS CONSISTENT WITH THE THEORY?

The key point is that a proper adaptation of the SRMC principle allows the above concerns to be taken into account, while the other procedures are much less adaptable and more awkward. But in order to derive these adaptations, a precise statement of the conditions for SRMC must be made.

As it is often the case in economic matters, the theory of charging is developed in a satisfying manner in the case of restrictive hypothesis describing a sort of perfect world, in which all markets are competitive, there is no external effect nor public good nor even state (or if there is a state collecting taxes, these taxes do not disturb the optimum). In that case, which will be examined first, the results are simple and clear, and provide a good link between infrastructure charging and investment choice. The formulas are simple and they emphasise the role of marginal cost as the rule for optimal charging.

Things become more complicated when departures from the first best optimum are taken into account. These departures involve various external effects, non-competitive markets, imperfect taxes have been analysed rather thoroughly by the theory and lead to formulas for charging which are just an extension of the principle of marginal cost charging. Equity concerns can also be addressed through clear and calculable formulae.

Things are more fuzzy when such features as uncertainty, information asymmetry or the institutional structure with separate jurisdictions are considered. Let us consider successively these three degrees of complication in the three following sections.

4.1. The First Best Situation

This first best situation is one where:

- there are no external effects (except, in the transport sector, the mutual annoyance that the users cause to each other), nor public good
- the firms behave in a competitive manner: they are price-takers

- there is no tax or taxes are optimal (essentially they are not linked to the economic activity of the agent; the typical optimal tax is the lump-sum tax)
- there are no information problems: the information is the same for anybody and there is no uncertainty
- there are no transaction costs
- there are no redistribution problems.

In that case, the general result is that marginal cost charging is the optimum and simple rules can be derived for the investment choice. This result is well ascertained, but the hypotheses of the first best case are not fulfilled:

- There are externalities and public goods
- Markets are monopolistic
- Taxes are not optimal
- Distribution is not optimal
- There are information imperfections (uncertainty, asymmetry)
- There are transaction costs.

Let us consider successively the four first imperfections, under the title of second best situations, where formulae for infrastructure charging can be expressed and written on the basis of SRMC, and the two last ones, the solution of which implies more qualitative answers.

4.2. Second Best Situations

Let us now consider several kinds of departure from the first best optimum pertaining to market imperfections but not to problems of information.

4.2.1. Externalities and Public Goods

Let us consider the case of environmental externalities: a well-known result is that it is then necessary to include the cost of environmental damages to the SRMC: this is the PPP.

4.2.2. Monopolistic Markets

Let us consider the case where the operator market, just downstream of the infrastructure market, is an oligopoly. The result is that the downstream market is priced above the marginal cost. In order to restore the optimum where the downstream market should be priced at the marginal cost, it is necessary to fix the infrastructure charge at a level lower than the SRMC; formulae can be easily derived for this situation.

4.2.3. Non-optimal Taxes

Let us consider, for instance, that an alternative transport mode is not priced at the right level, is under-priced; in that case, it is evident and this point can be demonstrated that the transport mode under examination should be also under-priced vis-à-vis the SRMC. Let us assume that we have two transport modes, named by the subscripts 1 and 2; let us assume that, for instance, for institutional reasons, the price of mode 2 is held different from the first best solution, namely that

$$p_2 \neq c_2$$

Then it is obvious that p_1 should be different from c_1 in the (second best) optimal situation. It can be shown (Quinet, 1998) that

$$x_1(p_1 - c_1) = -(e_{12}/e_{22})x_2(p_2 - c_2)$$

where e_{22} is the own elasticity of mode 2 and e_{12} the cross elasticity of mode 1 vis-à-vis the price of mode 2.

Another similar situation is the case where the general tax is not a lump-sum tax, but a tax based on the economic activity of the agents, for instance, the VAT. In that case, everything is as if a Euro raised for tax is more costly than a Euro raised for a private exchange: this is the cost of public funds. The optimal charge is then different from the SRMC, and is expressed through the well-known Ramsey-Boiteux formula (Ramsey, 1927; Boiteux, 1956, Appendix 2).

Let us note that in this situation, we implicitly assume that SRMC induces a deficit; this assumption is generally accepted, but probably not as evident as often thought. For instance, calculations of optimal pricing lead to the inverse conclusion: SRMC leads to surpluses and not to deficits, at least at a sufficient level of spatial aggregation (Roy, 2002).

4.2.4. Redistribution Concerns

The first best solution assumes that the distribution is optimal; it is possible to take into account redistribution objectives. The formula can be derived from the optimisation of users' utilities, weighting these utilities according to the preferences of redistribution.

4.2.5. The Limits of Partial Analysis. Advantages of a General Equilibrium Analysis

The previous results are generally obtained through partial analysis, considering the transport market without paying attention to the rest of the economy. It can be demonstrated that this procedure is valid only under limited hypotheses about the utility function, which should by quasi-linear; furthermore, the departures from the first best are taken separately.

If we want to depart from these hypotheses, it is necessary to use general equilibrium models with less exacting assumptions on the utility function. They allow income effects to be taken into account as long as the utility functions of the consumers have no restriction, and also to consider transport as an intermediate good.

It is also possible through these models to take into account imperfect competition, income distribution concerns (through a weighting of the utilities of the consumers), and tax imperfections. Boiteux (1956), Diamond & Mirrlees (1971), have explored the properties of such models when used for social optimisation purpose. Bovenberg and Van der Ploeg (1994) and finally Mayeres and Proost (2002) have introduced external effects in the model.

The most robust result is that the indirect tax in the sector of intermediate consumption should be zero in the absence of external effect; and, in the presence of external effects, it should be equal to the marginal external cost. The indirect taxes apply to the final demand addressed by the consumers; in an optimal situation, they are usually different from one good to the other. The formula of the optimal tax on transport includes the marginal costs of producing transport (infrastructure, congestion and environmental costs), the demand characteristics, especially the matrix of elasticities, and coefficients expressing the income distribution concerns. These formulas are generalisations of the previous formulas established under the assumption of partial analysis; but now they imply characteristics (prices, quantities, costs, etc.) of all markets, and not only of the transport market. These formulas will not be reproduced here.

4.3. Problems of Uncertainty, Information and Institutions

It then appears that, on the grounds of efficiency and equity, SRMC pricing – or formulas deduced from it in an 'imperfect' world – provides valid answers and leads to better situations than other concepts such as AC or LRMC. Let us now turn to more realistic situations of asymmetric information (for instance, the infrastructure manager has a better knowledge of costs and demand than the regulator), conflicting objectives between institutions (the regulator aims at welfare-maximising, and firms aim at profit-maximising), and where, furthermore, there is uncertainty (so that it is not possible afterwards to know whether a good result on costs is due to the productivity efforts of a firm or to chance).

It is clear that institutions are important for the analysis of these situations; for instance, the information asymmetry between infrastructure management and operations is much lower when both are included in the same organisation than when they are run by two different organisations. We assume that, in the line of the reforms of the European Union (and according to the real situation in most modes of transport), there is a separation between infrastructure management and the operators, and that the third actor is the regulator. The objective of the operators – which are generally private firms – is to make profit; the objective of the infrastructure manager is a mix of profit, bureaucratic goals (maximising the turn-over, etc.) and welfare maximisation, according to its status (rarely private firm, most often public firm or public service); the objective of the regulator is to achieve some kind of welfare maximum. In this framework, two types of relations and information situations are to analyse from the point of view of infrastructure charges: downstream, between the infrastructure manager and the operators, and upstream between the infrastructure manager and the regulator.

4.3.1. Relations between the Infrastructure Manager and the Operators

In this case, the important change is that neither the infrastructure manager nor the regulator know well the private value or social costs of the services to be run on the infrastructure. The challenge is then to induce efficiency in the downstream market of the operators. This opens the case for quantity regulation instead of price – and charges – regulation, and for the use of auctions instead of charges to select the efficient services. Let us see these points more in detail.

Prices or Quantities Regulations. On the side of costs, the uncertainty is large for environmental damages; nobody knows well the effects of car traffic in a specific situation on air pollution. In this situation, charges are very uncertain and the welfare gain is hazardous; quantity regulation may have more predictable and reliable effects (Weitzmann, 1974): according to the slopes of these two functions, and to the dispersions of the uncertainty on demand and on costs, it is better to achieve the allocation through prices or through quantities. In this last case, there is no charging, or charging is based on goals other than efficiency (for instance, according to the objective of revenue collecting), and efficiency goals are achieved through quantity regulation. Furthermore, a correct charging system would need a lot of information and implies transaction costs that would be too high for it to be valuable. This fact provides another argument in favour of quantity regulation.

The Use of Auctions. The efficiency of charges is that it is a mechanism for selecting the efficient services; this mechanism works well when the values of the operators are simple and not related to each other; it is the case in road transport: the individual users the value of which is higher than the charge are selected, the others are rejected, and the total value is maximised. This may not be the case in some modes such as rail and air transport, for instance, and more generally, when there is a need to allocate scarce capacity; in rail for instance, the value of a service for an operator depends on whether competitors have got another service close to the first one, or whether the first operator has got another service on a complementary line: the values of services are not independent and it can be easily shown that in such cases charging cannot select the most efficient operator, and charging must be replaced by auctions.

How to Induce Efficiency in Operations?. Another problem related to charging is to induce efficiency in operations. The solution depends on how the downstream market operates; if the downstream market is competitive, efficiency is implemented by competition itself, the infrastructure manager has not to deal with it, and the regulator has "just" to ensure that competition works well.

If the downstream market is not competitive, a possibility to achieve the goal of productivity or effort incentives to the users is to offer to the users a menu of tariffs (Laffont & Tirole, 1993), the simplest expression of them being often two-parts tariffs: a high quantity of transport is associated with a low unit price, close to the marginal cost, and will be chosen by efficient users, and at the other end, a low quantity associated to a high unitary price will be chosen by relatively inefficient users. The problem is that this situation may hamper new entrants which are small users, vis-à-vis the incumbent firms which are generally large users. This can be applied to any regulated transport production consumed by a segmented demand: travellers and freight on railroads or heavy goods vehicles and private cars on highways, or more simply high and low users of an infrastructure. The social surplus maximisation leads to the following best incentive to the regulated operator: high consumption should not be distorted (i.e. price should equal marginal cost), but low consumption is lower than the one when the operator has perfect information about users' preferences.

Well-devised tariffs can achieve this goal over time: when uncertainty is low, the charge is close to the "price-cap" type, which has a strong incentive effect and does not leave too much information rent; when uncertainty is high, charging is close to the type "cost plus": incentive must be lowered in order not to imply a too high information rent.

4.3.2. Relations between the Infrastructure Manager and the Regulator
The task to which charging can contribute is then for the regulator to induce efficiency in the infrastructure manager's behaviour. Efficiency must be taken in a very wide meaning: first to minimise the cost of infrastructure provision, then to ensure static efficiency (the infrastructure manager does not misuse its monopoly power), and also to achieve dynamic efficiency (mainly to induce a correct infrastructure investment).

The solutions – and their problems of implementation – are classical: regulation on infrastructure charges to avoid monopolistic behaviour, price-caps for cost-minimisation. A specific case of transport infrastructure is that it is believed to have large return to scale and fixed costs and may lead to rent-seeking on behalf of infrastructure managers whose objectives are more bureaucratic than welfare maximising: using the uncertainty on costs, they advocate low marginal costs, then low charges; they get large subsidies to cover fixed costs and large investment to cope for the demand, which is artificially increased by the low charge at the expense of the tax-payer. The argument is advocated by the supporters of average cost pricing. The solution of economic theory is to propose a menu of charges to the infrastructure manager, and it appears that the charge is higher – and closer to the average cost – the more uncertainty and information asymmetry is high.

Another argument advocated by the supporters of Average Cost Pricing is that without a link between expenses and revenues, and a punishment if expenses exceed revenues, there is a risk of over-investment due to the uncertainty of investment appraisal and strategic behaviour by infrastructure managers who are guided by bureaucratic goals of extension in size of their business. The criterion of Average Cost Pricing ensures that the investment is profitable, but it leads to reject other investments which are also desirable but do not satisfy this criterion. Let us also assume that the main idea underlying this question is to put a link between expenses and revenues, but this link may not be just an equality; a well-devised lump-sum subsidy may both achieve this goal and allow for a SRMC charging. Anyway this situation calls for an improvement in the expertise in project appraisal.

The problem of implementation of this kind of solution is the risk of regulatory capture: as the time goes, there are more and more renegotiations of the contract between regulator and infrastructure manager, the regulation becomes more and more discretionary, and the risk of regulatory capture increases. This argument calls for a deeper attention to institutional arrangements. A solution could be that the infrastructure manager has to break-even – if it is a public firm – or to maximise its profit under price regulation – if it is a private firm – with lump-sum subsidies linked to fixed costs estimates. Another point which implies institutional arrangements is the situation when there are specific investments: then a strong partnership or a merger of infrastructure managers and operators is to be recommended. In this situation, the infrastructure charging problem vanishes.

Specific attention has to be given to the fine-tuning of the financial arrangements. For instance, a private infrastructure manager whose objective is profit should not be directly given the revenues from congestion costs as this would induce him to under invest in capacity infrastructure; these revenues should better go to him through lump-sum subsidies.

4.3.3. Problems of Jurisdictions

Problems of jurisdictions arise first in the case of international traffic. Such a situation has been analysed, for instance, in the case of international traffic between two countries, this traffic being charged in both countries for the part of the trip it makes in each of them (Courcelles, de Borger, & Swysen, 1998; Bassanini & Pouyet, 2000).

The pricing in each country has effects on the tax revenue and on the externalities in the other country. The pricing policy depends on the degree of co-operation or competition between the two countries, and on the technical and legal possibility to discriminate between domestic and foreign traffic; it depends also on the hierarchy of environmental goals and budget constraints.

Another point which begins to be addressed by economic analysis is the effect of the hierarchy of jurisdictions. Reasons in favour of such hierarchy are the cost of information transmission and processing, incompleteness of contracts between the centre and the periphery, and a better effort incentive (Caillaud, Jullien, & Picard, 1996); on the other side, taxes by inferior jurisdictions are subjects to several drawbacks, especially in the case of source-based taxation: tax exportation, competitive tax spill over, NIMBY and beggar-my-neighbour tax competition induces inefficiencies in state taxes; this point leads to recommend to use resident based taxes by the central government in order to correct these failures (Inman & Rubinfeld, 1996).

Another point is the impact of practical implementation of charging. Though the optimal pricing is very diversified (according to the time, the location, the type of vehicle), this practical implementation leads to averaging the categories, and averaging the categories automatically leads to a reduction in welfare. This point argues for high differentiation; the argument against high differentiation is, of course, the transaction cost of differentiation. Very little knowledge is available about this trade-off. It would be useful to achieve calculations on the magnitude of the effects of differentiation, based on some practical examples (for instance, differentiation against averaging in the case of heavy goods vehicles).

4.3.4. Conclusions on Information, Uncertainty and Institutions

What can be concluded from these considerations on information, uncertainty and institutions?

- First, if problems of information and uncertainty are not taken into consideration, then economic theory has precise answers to the problems arising from the various imperfections of the real-world vis-à-vis the hypotheses of an ideal (first-best) economic world. These solutions are based on SRMC, and depart from it through well-known formulas.
- When information and uncertainty problems are taken into account, things are more complicated. In some cases, the solutions do not rely on charges, but on regulations or some revelation mechanism such as auctions. In almost all these situations, the charging principle must be set up in accordance with – and closely linked to – institutional arrangements.
- The decision to take depends on how large are the fixed costs and the information asymmetries.
- Are fixed costs high? Econometric analyses show that fixed costs are not that high and that they have been often overestimated in the past. Often, fixed costs are due to the fact that infrastructures are over-dimensioned, either because they have been provided for public obligation purposes (rural roads) or because traffic forecasts were wrong. In the first case, the charging problem is mainly a matter of political choice and equity; in the second one, the solution is to close the infrastructure, and is not a charging problem, except if some kind of charging may induce this closure and does not entail too many drawbacks on other grounds.
- If fixed costs are high, the infrastructure is then more akin to a pure public good. In this last case, the situation clearly advocates for public management, i.e. integration between the infrastructure manager and the regulator. The same happens when public service obligations and external effects are very important. The charging issue is then much less important than the institutional issue.

- Is information asymmetry high? The answer is highly depending on specific situations. If it is the case, a budget balance solution is necessary, but it does not prevent an SRMC-based charging solution, of the Ramsey type.
- In some specific case, charging must be replaced by regulations or by auctions.

5. OVERALL CONCLUSIONS

On the whole, it appears that there are large discrepancies about charging doctrines, on the fields of academic teaching (advanced courses teach SRMC principle and its limitations, other teach accounting-based procedures), of official doctrines (some states support average cost principle, other ones are closer to SRMC principle), and of practical implementation (in most modes, real charging systems more or less tend to achieve break-even).

These results can be viewed through a positive analysis, looking at the behaviours of decision-makers, or through a normative analysis, looking whether the economic analysis has answers to their concerns.

It appears that political decision-makers are more interested in acceptability and equity than in efficiency, and in fact, SRMC often implies unfavourable consequences on these grounds. Furthermore, SRMC appears to be a complicated concept, it must be acknowledged that its estimates are uncertain and this uncertainty leads to strategic behaviour by the operators, the result being increases of deficits, subsidies and costs.

Alternative concepts are suggested by decision-makers in order to overcome these drawbacks. They are AC, LRMC or DC. What are the assessments of economic analysis on these questions?

Economic analysis puts more emphasis on efficiency. On this ground, a first limitation of SRMC pricing arises from the fact that there is no practical tool to implement the very diversified charges of the SRMC, and proxies imply loss of efficiency. A second limitation comes from the fact that the assumptions of the first best world are not fulfilled: apart from information problems, there are market imperfections (non-competitive markets, cost of public funds, externalities). Economic analysis provides answers to these situations. It provides also formulas to solve equity problems. Those formulas are derived from SRMC, they include information on demand (mainly elasticities). Without information and uncertainty, these formulas are superior to the alternative doctrines such as AC or DC, as they are the optimal solution provided by economic theory in the framework of general equilibrium models.

Things change a lot with uncertainty and information asymmetry. It appears then that charging may not be the proper way to achieve efficiency: in some cases, regulations or auctions should be preferred. Furthermore, alternative doctrines gain in interest, as they prevent strategic behaviour such as rent-seeking and they induce more incentives for cost minimisation.

The issue is: how large are fixed costs and information asymmetries:

- Recent econometric analyses show that fixed costs have been often overestimated in the past and are not that high. Often, fixed costs are due to the fact that infrastructures are over-dimensioned. In this case, the charging problem is mainly a matter of political choice and equity. The solution is to close the infrastructure, and is not a charging problem, except if some kind of charging may induce this closure and does not entail too many drawbacks on other grounds.
- If fixed costs are high, the infrastructure is then more akin to a pure public good. The solution is not to be found in charging but in the institutions which allow for better decisions; clearly, this situation advocates public management, i.e. integration between the infrastructure manager and the regulator. The same happens when public service obligations and external effects are very important.
- Is information asymmetry high? The answer is highly depending on specific situations. If it is the case, a budget balance solution is necessary, but it does not prevent a generalised SRMC charging solution, of the Ramsey type.

Anyway, whatever the charging solution is, its implementation needs to know the cost function of the infrastructure, an approximation of which is given by the knowledge of the SRMC. Furthermore, in almost all situations, the charging principle must be set up in accordance with and closely linked to institutional arrangements.

It is then important to assess, in each specific situation, how far are we from the hypotheses of first best, how large are uncertainty and information asymmetry, and how large are fixed costs. The opinion of the author is that, on those matters, we are in general not far from first best hypotheses, that fixed costs are not that high, and that consequently there is a large scope for SRMC pricing.

But it should be acknowledged that infrastructure charges are just tools to achieve goals, and that these goals can in many circumstances be reached through other tools; for instance, regulations may be preferred to charges when uncertainty is high, as well as permit markets or auctions in the case of strong information asymmetries. Furthermore, the institutional arrangements may lead to strong diversions from the initial goals of the charging system; for instance, a marginal cost pricing system provides a strong incentive to the infrastructure manager to adopt a rent-seeking behaviour and to try to lower the marginal costs; these examples remind us that the consequences of a charging system must not be considered per se, but along with its institutional and sociological environment.

NOTES

1. G.N. Merchan (EIET) for Spain, E. Quinet (CERAS) for France, T. Sansom (ITS) for Ireland and United Kingdom, S. Suter (Ecoplan) for Austria, Germany and Switzerland.

2. We will not try to put a rigorous definition of equity; let us just say that there are two main strands in the literature: the social welfare function approach that uses a weighed sum of changes in real income and cope with distributions of income; and the approach that starts from specific moral judgements, for instance in the line of the distributive justice of Rawls.

REFERENCES

Bassanini, A., & Pouyet, J. (2000). *Access pricing for interconnected vertically separated industries*. Communication to the Helsinki Workshop on infrastructure charging on railways.

Boiteux, M. (1956). Sur la tarification des monopoles publics astreints à l'équilibre budgétaire. *Econométrica, 24*, 22–40.

Bovenberg, A. L., & Van der Ploeg, F. (1994). Environmental policy, public finance and labour market in a second best world. *Journal of Public Economics, 55*, 349.

Caillaud, B., Jullien, B., & Picard, P. (1996). Hierarchical organisation and incentives. *European Economic Review, 40*, 687–695.

Castano-Pardo, A., & Garcia-Diaz, A. (1995). Highway cost allocation: An application of the theory of non-atomic games. *Transportation Research A, 29*(3), 173–186.

Commission of the European Communities (1998). *Fair payment for infrastructure use: A phased approach to a common transport infrastructure charging framework in the EU*. White Paper, COM (1998) 466 final. Brussels: European Commission.

Courcelles, B., de Borger, B., & Swysen, D. (1998). *Optimal pricing of transport externalities in a federal system: A theoretical analysis*. Communication to WTCR (Antwerpen).

Curien, N., & Gensollen, M. (1992). *L'économie des télécommunications: Ouverture et réglementation*. Paris: Economica-ENSPTT.

Diamond, P., & Mirrlees, J. Optimal taxation and public production, parts I and II. *American Economic Review, 61* (March, June 1971), 8–27, 261–278.

Highway Research Board (1962). *The AASHO road test: Pavement research*. HRB Special Report 61E. Washington, DC.
Inman, R. P., & Rubinfeld, D. L. (1996). Designing tax policy in federalist economies: An overview. *Journal of Public Economics*, *60*(3), 307–334.
Laffont, J.-J. (1984). *Théorie micro-économique*. Paris: Economica.
Laffont, J.-J., & Tirole, J. (1993). *A theory of procurement and regulation*. Cambridge: MIT Press.
Link, H., Stewart, L., Maibach, M., Sansom, T. & Nellthorp, J. (2000). *The accounts approach. UNITE (UNIfication of accounts and marginal costs for Transport Efficiency) deliverable 2*. Funded by 5th Framework RTD Programme. ITS, University of Leeds, Leeds, October.
Mayeres, I., & Proost, S. (2002). Reforming transport pricing: An economic perspective on equity, efficiency and acceptability. Communication to the 4th IMPRINT Seminar, Brussels.
Peter, B. (2004). *Rail infrastructure: Pricing and investment*. Workgroup on Infrastructure Policy.
Quinet, E. (1998). *Principes d'économie des transports*. Paris: Economica.
Ramsey, E. (1927). A contribution to the theory of taxation. *Economic Journal*, *37*(March), 47–61.
Roy, R. (2002). *The fiscal impact of marginal cost pricing: The spectre of deficits or an embarrassment of riches?* Communication to the 2nd IMPRINT Seminar, Brussels.
Weitzmann, M. L. (1974). Prices versus quantities. *Review of Economic Studies*, *41*, 477–491.

APPENDIX

The details of the answers to the survey are given in Tables A1–A3 and reflect the position in 2001.

Table A1. What are the Differences between the Picture given by the Benchmark Doctrine and the Current Teaching at Universities about Transport?.

Country	Answer
Austria	Marginal cost pricing is taught in the context of microeconomics but is not considered as a possible implementation principle. Pricing has been discussed first as a funding generator.
France	Economic theory is taught in the more advanced economics courses in universities; but in other courses (equivalent to MBA) less sophisticated methods are taught; they are based on principles of cost allocation.
Germany	Marginal cost pricing is considered as a theoretically interesting approach but not as an important input for transport pricing in practice. Comments on the White Paper on Infrastructure Charging (CEC, 1998) were very critical from the academic world as well as from representatives of the relevant parts of the public administration.
Ireland	Not known. Transport economics is not widely taught.
Spain	Students generally are shown the main principles of economics theory. But most of transport courses in Spain are more often offered by engineering schools and tend to stress more the technical analysis.
Switzerland	Transport economics is not widely taught. The two national technical universities in Zurich and Lausanne offer courses in transport science, but the approach is rather engineering and planning than economics. In the last years, transport economics has been the subject of two National Research Programmes, which included research on the question of different pricing approaches in transport.
United Kingdom	Advanced theoretical courses cover classical economic theory, but there is still a tendency to teach traditional cost allocation procedures.

Source: CEC (1998).

Table A2. What are the Current Doctrines Expressed by the Political Authorities (Government, Parliament, etc.) on the Subject of Transport Infrastructure Pricing?.

Country	Answer
Austria	The priority objective of environmental protection was implemented through regulatory and pricing measures. Nevertheless, pricing measures introduced so far serve first of all for the generation of funds for the general budget and the financing of the transport infrastructure, though a on-going project on road transport infrastructure costs will most probably result in an opening of the discussion about this issue.
France	The doctrine has varied over the years. About 20 years ago, the principle was that freight should pay the marginal cost, and passenger traffic should pay the full cost. More recently, the main stream of ideas shifted towards the use of LRMC principle, based on concerns about the manipulability of SRMC and on (inter-modal) equity considerations.
Germany	The current pricing doctrine is dominated by financing issues and not by considerations referring to marginal cost pricing. The discussion on environmental taxation relates more easily to marginal cost pricing.
Ireland	There is no move for pricing of inter-urban road networks (with the exception of tolled bridges, for the purposes of project finance). There is no pressure for road pricing in Dublin, although studies have been commissioned in the past (e.g. with a view to developing finance sources for light rail). For other sectors, there is no political momentum behind changes in charging policy.
Spain	The previous administration launched plans based on publicly financed investments. After 1996, the new government has shifted the balance slightly towards a model of charging infrastructure costs to users.
Switzerland	Recently, it has become clear that SRMC pricing is considered as an interesting economic approach but not as central future guideline. In practical transport policy, short-run efficiency is not considered as a very important objective of pricing in comparison with financing and environmental objectives. Environmental costs play a role in pricing policy. However, the basic idea of marginal cost pricing is not considered as feasible because of uncertainties in the calculation of marginal costs. SRMC pricing has become an issue within the context of railway reform (as a baseline for track pricing).
United Kingdom	There is a tradition going back to the 1960s in favour of LRMC pricing, combined with a current strong encouragement towards congestion pricing for both road (delegated to local authorities for urban roads) and rail, which may be taken to indicate a move towards SRMCs. There is a minimal interest in charging issues in the ports, aviation or inland waterways sectors.

Table A3. What is the Real Situation of Present Infrastructure Pricing?.

Austria	Road: The taxation of fuel is not earmarked. There is a purchase tax and an annual vehicle tax. At the local level, there are parking fees. Passengers cars and light goods vehicles (< 12 t) have to buy a vignette to use the motorways. On five roads and several tunnels there are road tolls. In addition to the tolls, heavy goods vehicles > 12 t pay an annual road user charge ("STRABA"). Furthermore, an eco point system for transit traffic through Austria exists. A distance-related charge ("Maut") for vehicles > 3.5 t is planned by 2002 on motorways and other trunk roads for funding the extension and maintenance of the high-ranking road network, operated by a state-owned company. Rail: The infrastructure access charge is a tariff based on two variable parameters: train-km and gross-ton-km. It is not based on marginal cost estimates. Air and inland waterways: The level of the charge is derived from total cost estimates and not from marginal cost considerations.
France	Road: Road is charged through many devices: fuel taxes, toll motorways, vignettes, parking fees. Though not determined by the same authority, their main motivations are financial and not economic. The outcome is that, roughly, road as a whole covers its charges, but with a lot of discrepancies between categories of traffic. Rail: Rail transport is subsidised, the infrastructure charges cover about 25% of the total expenses. Charges approximately follow the Ramsey-Boiteux principle. Air and sea transport: They roughly pay their expenses, as they are run by (public) firms and do not receive much subsidy from public authorities.
Germany	Road: The main pricing scheme is the taxation of fuel. An annual vehicle tax is levied by the different states in Germany. At the local level parking fees are levied. Heavy goods vehicles using the German motorways pay for the Eurovignette. Only part of the revenues from the duties on fuel is earmarked for the financing of road infrastructure costs. For the short to medium term the approach is to base the financing of infrastructure more on user and less on budgetary funding. Distance-related user charges should be introduced on motorways for heavy goods vehicles (> 12 t), then extended to other road types. The introduction of road pricing for cars has been rejected. The level of the user charge will be derived from estimates of total and not marginal infrastructure costs. In the short- or medium-term buses and light goods vehicles will have to buy a time-dependent vignette to use the German motorways. For road passengers cars a motorway vignette is in discussion. Rail: In 1998 a two-part tariff system of infrastructure access charges was introduced. It shows similarities to a marginal cost pricing scheme subject to a budget constraint. However, the German cartel office rejected this pricing system. A new system will have to be elaborated.

Table A3. *(Continued)*

Ireland	Road: Generally uncharged. Rail: User tariffs have been determined over time, generally maintaining parity with bus and coach services. Infrastructure cost coverage has not been sought, nor has any specific infrastructure pricing policy been developed.
Spain	Road: There are no developed pollution or congestion charges, and vehicles pay for the use of roads through taxes on fuel, annual licenses and other charges. Rail: The public railway company is in a process of transformation towards a model of separation, but it is not clear yet what are the plans for the agency that in the future will be in charge of managing infrastructure. Seaport and airports: They are generally self-financed through their revenues, so for those modes users cover for infrastructure costs.
Switzerland	Road: The main pricing instruments are: fuel taxes (whose revenues are partly earmarked for the financing of road infrastructure), annual vehicle taxes (levied at the cantonal level), parking fees (levied at the local level), an annual earmarked vignette on cars for the use of the national motorways. A new Heavy Vehicle Fee will be introduced in 2001 the level of which depends on truck characteristics (weight, emission technology, etc.) and on the need to finance rail investments. Rail: The charge should not be lower than the marginal cost incurred with the use of a "standard" part of the network. In addition, a contribution margin can be levied to contribute to cost recovery. Inland waterways: Relevant are only the charges levied in the Rhine harbours of Basle. They are not based on any marginal cost estimates. Airports: Landing charges are oriented at financial considerations and include environmental considerations (the noise emissions of the aircraft).
United Kingdom	Road: Annual vehicle licence duty and fuel tax; there is no explicit link between these and costs although relative damage estimates are used to establish differentials. Rail: Infrastructure charges for franchised passenger operators are based on a two-part tariff; the rail regulator has recently allowed Railtrack to increase the variable part of the tariff bringing it more into line with marginal cost, but maintaining revenue neutrality. Airports and seaports: There is minimal government involvement in charge setting. There is no regulation of port prices, and ports are generally under private ownership in a competitive marketplace. The Civil Aviation Authority regulates airport charges for major airports, with an emphasis on infrastructure cost recovery rather than the application of economic principles for charging.

INFRASTRUCTURE

Heike Link and Jan-Eric Nilsson

1. INTRODUCTION

If additional vehicles affect the need for infrastructure maintenance, repair, renewal and operation, a welfare-maximising infrastructure policy should take this into account through its charging scheme. While extensive studies on optimal congestion and environmental charges are available, less attention has been paid to the estimation of marginal infrastructure costs. One reason may be the presumed lesser importance of this cost component; environmental and congestion costs are believed to be decisive for the overall level of charges.

The purpose of this chapter is to address issues of principle and methodological relevance for measuring the costs of marginal infrastructure use. Although we also summarise empirical evidence in this field, it should be emphasised that the methods – and indeed the resulting estimations – that are presented may well be obsolete within a few years due to the fact that the relevant research frontier seems to be moving quickly.

There are two very practical questions that come back over and again when analysts seek to understand the way in which costs for infrastructure use vary. The first is related to the distinction between wear- and weather-related costs and the consequent fact that much maintenance work would have had to be dealt with even without traffic. The policy issue is then how much revenue a price tariff based on marginal costs for infrastructure wear would raise; except for this, charges would also include a component to account for scarcity which is not the target of the present chapter. The second question relates to which classes of costs are to be included in the

Measuring the Marginal Social Cost of Transport
Research in Transportation Economics, Volume 14, 49–83
Copyright © 2005 by Elsevier Ltd.
All rights of reproduction in any form reserved
ISSN: 0739-8859/doi:10.1016/S0739-8859(05)14003-7

analysis: Should both costs for operating the facilities, for maintaining and for renewing them when that day arrives, be made part of the analysis? Both questions must be addressed using both analytical reasoning and looking at data. As will become apparent from our review, data-capture is at present difficult, and analysts often have to start with making assumptions and testing hypotheses against poor evidence to get anywhere.

This chapter is organised as follows. Section 2 will discuss the main methodological approaches available for estimating marginal infrastructure costs. Section 3 deals with methodological issues and empirical results per mode of transport. Section 4 draws conclusions for transport policy and future research. Since less research seems to exist about the possible marginal wear-and-tear costs for using airports and seaports, inland waterways, public transport infrastructure and railway stations, this is not dealt with in the review.

2. METHODOLOGICAL ISSUES IN MEASURING MARGINAL INFRASTRUCTURE COSTS

The estimation of infrastructure cost functions makes use of two classes of methods. Econometric approaches – further presented in Section 2.1 – make use of observed expenditures on infrastructure maintenance, repair and operation and observations of (possible) cost drivers, either based on cross-sectional or longitudinal data. The obvious point of departure for this research is the existing literature on cost functions in general. Much of the previous work has been motivated by analytical problems other than estimating infrastructure costs, notably deregulation issues or attempts to measure productive efficiency across firms over time (for example Caves et al., 1985; Bauer, 1990; Grabowski & Mehdian, 1990; Talvitie & Sikow, 1992). These studies focus on transport companies (trucking and rail companies and airlines). Transport infrastructure is part of the cost estimations for operating (U.S.) vertically integrated railways, while typically not for other modes.

The second technique will be referred to as engineering-based approaches, deriving functional relationships from measurements of infrastructure damages and cost drivers and evaluating these relationships in monetary terms (Section 2.2).

A further useful distinction is that between using bottom-up and top-down inference. Bottom-up approaches consider in a first step the costs of so-called basic packages, for example, costs for constructing a piece of infrastructure for the least demanding vehicle category. In a stepwise approach the additional costs caused by successive vehicle categories that require more of the infrastructure are added. If these categories are defined in a sufficiently detailed way, the bottom-up approach could be considered as a discrete (or incremental) approach to the first derivative of a cost function, e.g. to the marginal cost function. In contrast, top-down approaches start from observed total costs, possibly accounted for on a piece-meal way.

A further type of research in the field of infrastructure costing are infrastructure cost studies in Germany, Austria, Switzerland and France, which have been aimed at quantifying the total costs of infrastructure and allocating them to different users or vehicle types. These studies have been motivated by a general interest in cost information. To our knowledge, no rigorous cost function analysis has been performed on such data. However, some of these studies have attempted to distinguish between fixed and variable costs and to allocate top-down percentages of variable costs to different vehicle categories based on empirical, engineering and expert judgement; more on this in Section 2.3.

2.1. Econometric Approaches

Although a rich body of literature on production and cost functions is available, econometric cost function studies purely for infrastructure networks and terminals are rare. Early examples of cost studies are Lardner (1850), Lorenz (1916) and Ripley (1923), dealing with the question of cost variability, the problem of allocating joint and common costs and making attempts to measure returns to scale. All of them related to the rail industry, but similar questions were subsequently analysed for other modes too. While the early works used simple linear cost functions, the succeeding research made use of Cobb–Douglas production functions (see, for example, Keeler, 1974) and in a next step, flexible functional forms such as the generalised Leontief function and the translog function (Christensen, Jorgenson, & Lau, 1975). Berndt and Khaled (1979) proposed a generalised Box–Cox specification, which includes both the generalised Leontief function and the translog function as special cases.

This stream of literature experienced a new stimulus with the advent of contestability theory and the conceptual foundations for multi-product cost functions including scope economies. Another stream of cost function research was motivated by the question of measuring and comparing productive efficiency across firms over time by means of firm and time effect models (Caves et al., 1985) and by using frontier cost functions (Bauer, 1990; Grabowski & Mehdian, 1990; Talvitie & Sikow, 1992).

Of all functional forms tested over the last 30 years for describing producers' cost behaviour, the translog approach is the one most frequently used (for an overview of applications in transport, see Oum & Waters, 1998). It imposes few restrictions on the underlying production technology, allows a simple computation of substitution elasticities and contains all relevant properties of neoclassical production theory such as factor substitution, economies of scale and technological change.

The translog cost function describes the relationship between the cost C, an m-dimensional vector of output levels $\mathbf{Y}$ and an n-dimensional vector of input prices $\mathbf{W}$ by the logarithmic function

$$\begin{aligned} \ln C = a_0 &+ \sum_{i=1}^{m} a_i \ln Y_i + \sum_{j=1}^{n} b_j \ln W_j \\ &+ \frac{1}{2}\sum_{i=1}^{m}\sum_{j=1}^{m} a_{ij}(\ln Y_i \ln Y_j) + \frac{1}{2}\sum_{i=1}^{n}\sum_{j=1}^{n} b_{ij}(\ln W_i \ln W_j) \\ &+ \sum_{i=1}^{m}\sum_{k=1}^{n} d_{ik}(\ln Y_i \ln W_k) \end{aligned} \tag{3.1}$$

Duality theory states that under certain conditions C is dual to the transformation function $T(\mathbf{Y},\mathbf{X})$ where $\mathbf{X}$ is an n-dimensional vector of factor inputs. Symmetry conditions for the second-order coefficients are imposed. Normally, the translog function is estimated jointly with the cost-minimising input cost share functions by means of seemingly unrelated regression (SUR). The cost-minimising factor demands can be obtained by applying Shephard's (1970) lemma

$$s_i = \frac{W'_i X_i}{C} = \frac{\partial C}{\partial W_i}\frac{W_i}{C} = \frac{\partial \ln C}{\partial \ln W_i} \tag{3.2}$$

and must sum to 1. The first row of Eq. (3.1) gives the special case of a Cobb–Douglas function provided that all second-order parameters are zero. The following rows specify the cross-relationships within the output and input vectors themselves and amongst each other.

This traditional way of analysing how cost, output, production factors and input prices vary is also applicable for analysing the spending behaviour on infrastructure maintenance, repair, renewal and operation. However, further important influence factors such as climate, the number of years passed after last renewal of a motorway or rail section, the initial construction standard of infrastructure, etc. have to be considered in addition.

2.1.1. Output Definition and Separation

Special attention is required for defining the output vector **Y**. The immediate way to define output of infrastructure maintenance, renewal or construction would be to measure the volume of maintained, renewed or constructed road or rail track, expressed for example, as km, m^2 or m^3. An example of such an analysis is given by Talvitie and Sikow (1992) who seek to estimate productivity in highway construction, and therefore define the volume of construction work, measured in m^3, as output. The focus on productivity measurement does, however, mean that aspects related to allocative efficiency are neglected. The way in which infrastructure is dealt with could be highly (cost) efficient, but if the roads or tracks are not used by traffic, this would be of lesser overall interest.

Existing cost function studies of the trucking industry or for rail transport (e.g. Daughety, Nelson, & Vigdor, 1985; De Borger, 1992; Cantos, 2000) also make use of train-km, gross-tonne-km or passenger-km. These outputs are, however, produced by a rail operator or a haulage company with a combination of factor inputs such as energy, material, labour and capital, infrastructure being complementary only.

For studies of the marginal cost of infrastructure use, it is necessary to measure the output of infrastructure provision as the traffic making use of the facilities, with their respective second-order and interaction terms in a cost function approach. Link (2004, to be reviewed below) however estimates two separate models, one with an output vector defined as m^2 renewed road, and another one where traffic volume is defined as output.

Considering traffic volume as output vector leads to a specific problem of output separation. Normally, different classes of vehicles make use of transport infrastructure. Each of them may have a different impact on infrastructure damage and may also require different access to capacity. Difference in axle weight across road vehicles or in the types of tyres used may affect road wear in different ways and, similarly, different classes of rail traffic – passenger and freight or high- versus conventional-speed trains – may wear down the infrastructure in different ways. This would make it necessary to account separately for different classes of output according to well-defined criteria.

Such detailed information is often difficult to get access to and the standard approach has become to use some aggregate, such as gross tonnes passing over different sections of the infrastructure. The specific use of traffic volume data can also lead to various estimation problems. If, for example, an attempt to model road maintenance costs contains information about mileages of different vehicle types as explanatory variables, the results might be characterised by considerable multicollinearity problems. The consequence is either insignificant or unreliable coefficients (as reported by Herry & Sedlacek, 2002; Schreyer, Schimdt, & Maibach, 2002). Herry and Sedlacek (2002) and Schreyer et al. (2002) therefore *constructed* this type of information from traffic models based on data from (only few[1]) Weight-in Motion (WIM) stations and adjusted the data by growth factors to the years before and after the base year of the traffic model. Although it is preferable to have access to original data on gross-tonne or axle-load km, it should be acknowledged that tricks of this nature may be necessary in order to provide at least some insights into the nature of cost structures.

Another approach to the problem with multiple outputs is to estimate separate models for each type of vehicle; this is done in Schreyer et al. (2002). The problem here is that the models do not reflect the interaction of different vehicle types in their joint impact on infrastructure damage. Furthermore, these models usually do not achieve a high explanatory power.

An additional problem can occur if different traffic variables are used and some of them enter the model with zero observations with non-defined logarithms (for example, at certain rail sections only passenger or freight trains are operated). This can be handled by using Box–Cox transformations for these variables $(y_i^{\lambda} - 1)\lambda$ as proposed by Berndt and Khaled (1979) and by de Borger (1992). An example for a recent cost function study on rail track maintenance costs in France which uses a Box–Cox approach is Gaudry and Quinet (2003). Greene (2000) suggests to test a Box–Cox model specification before estimating a non-linear unrestricted Box–Cox model by means of a Lagrange multiplier test (see also Davidson & MacKinnon, 1985). Applying this test might of course lead to rejection of the hypothesis of a Box–Cox model formulation and thus to the exclusion of all zero observations.

2.1.2. Input Prices

The standard formulation of a cost function includes a vector of input prices as explanatory variables and the input cost share equations obtained by Shephard's (1970) lemma (see Eqs. (3.1) and (3.2)). Neither Johansson and Nilsson (2004) nor Gaudry and Quinet (2003), which applied a translog cost function and a Box–Cox cost function, respectively to analyse rail track maintenance costs, use such data. Both studies argue that differences in input prices across observations are negligible. Even if such prices could be readily obtained, it would therefore not affect the outcome of the analysis. Johansson and Nilsson (2004) state that input prices are largely harmonised within Finland and Sweden where two cross-section comparisons were made. Gaudry and Quinet (2003) argue in the same direction claiming that worker's wages are uniform across the national rail carrier SNCF (Société nationale chemins de fer France) and that other input prices differ not much among the French regions. Both studies therefore estimate only the main equation of the traditional cost function but not the input cost share equations.

The major issue in considering input prices in cross-sectional cost functions is, again, data availability. In particular, for cross-sectional data (Schreyer et al., 2002 analysed data with a mean section length of 15 km, Link (2004) used data with a mean section length of 7 km), it is difficult to obtain data on labour, material and capital prices.

2.1.3. Time-Series-based Approaches

While most econometric cost studies have used cross-section data or longitudinal data that have been aggregated, a complementary methodological approach – in particular for analysing operating costs of terminal infrastructure (airports, seaports) – is to use multivariate time series analysis with short observation units such as hours. Himanen, Idström, Göbel, and Link (2002) provide an example. Their analysis focused on labour costs as the dominant cost component for an airport and sought to identify the relationship between labour costs, aircraft movements and number of passengers in an hourly pattern. The study makes use of simple linear equations which seem to provide plausible cost estimates; they do, however, report problems with autocorrelated residuals. The obvious conclusion from this is to apply more advanced time series models, which also seek to consider external shocks (such as aircraft delays).

2.2. Engineering Approaches for Marginal Cost Estimation

In contrast to econometric cost function modelling, engineering methods are usually bottom-up approaches based on measuring physical relationships between damages of infrastructure and impact factors such as infrastructure use, climate etc. The most renowned example of this stream of research is the American Association of State Highway Officials (AASHO) road test (see Highway Research Board, 1961), which used trucks driven over roads built specifically for this purpose to derive a relationship between road damage and axle weight. The so-called fourth power rule, which emanates from these trials, indicates that road damage is proportionate to axle weight raised to the power 4. In other words, doubling the axle weight increases road damages by a factor of 16. This result leads to the conclusion that damage to a road is proportional to the number of standard axles passing over it, which is measured as the ratio of the sum of actual axle weights raised to the power 4 over some predefined standard axle weight (usually 10 tonnes) raised to the power 4. If road infrastructure costs are then assumed to be proportional to road damages, this damage function can be translated into a cost function.

An example of a modelling approach of the same nature is the MARPAS model applied by Railtrack, the U.K. rail track provider, for calculating maintenance requirements of rail infrastructure. Engineering-based approaches often use simulation techniques, where experimentally derived relationships between infrastructure wear and traffic are applied to bring order into the financial data at hand. An example is Wechsler (1998) which tests the results of the AASHO road test with traffic data from Germany and concludes that the fourth power is in principle applicable, however, with a range of 2.5–6. Small, Winston, and Evans (1989) report another re-examination of the original AASHO data, again concluding that the relevant power varies with the thickness of the pavement.

None of these engineering studies have been designed in order to estimate marginal costs. Their purpose is rather to provide indications of maintenance and renewal needs on a disaggregated level (for road or rail sections, etc.). Such needs are originally expressed in physical terms, for example, in terms of terminal values for road condition scores or in terms of values for the asset's lifetime, and evaluated with unit costs, average costs for necessary maintenance and renewal measures per km or m^3. Lindberg (2002) goes a step further and derives marginal costs for Swedish roads per standard axle and for different road construction characteristics and vehicle types; his focus is, however, on renewal costs – in particular resurfacing – rather than annual maintenance. The analysis makes use of engineering data to estimate a so-called deterioration elasticity of road surface and average costs per renewal measure (for more details see Section 3.1).

In contrast to econometric approaches that reflect *actual* spending behaviour, marginal costs derived with engineering approaches may reflect maintenance and renewal *needs*. They can thus be characterised as "ideal world" approaches which assume that maintenance and renewal practice always follow technical needs, an assumption which may be violated. Charging marginal costs that have been derived in this way may therefore result in overcharging, relative to actual expenditure.

2.3. Distinction between Variable and Fixed Costs in Top-Down Costing Studies

For the road sector, several national cost allocation studies that distinguish between fixed and variable costs are available. These allocate top-down percentages of costs that are assumed to vary with traffic volume to different cost categories based on empirical, engineering and expert judgement. An overview of such studies in the field of road transport is given in Link, Dodgson, Maibach, and Herry (1999) (see Table 1 for a summary). Such studies can be seen as the first step in estimating cost functions. By dividing total costs into fixed and variable costs, they provide some intuitive background information about the relevant part of total costs to be included in the estimation of marginal costs. An assumption often applied is that average variable costs (AVCs) are equal to marginal costs in order to have a simplified procedure to obtain the pricing-relevant marginal costs. The underlying assumption of this is that the costs for taking care of the infrastructure varies linearly with traffic volume, an assumption which is not proven, and which is in contradiction to results from engineering experiments (ASSHO road test).

For the rail sector, the national rail companies should have access to detailed accounts and information on variable and fixed costs. Booz, Allen, and Hamilton (1999) is, however, one of the few examples from the railway industry of top-down studies of costs. The authors review international research on use dependent track costs; results seem to indicate that between 30% and 60% of track maintenance and renewal costs vary with track use. We do not know of such accounting studies for other types of infrastructure such as seaports, inland waterway and airports.

3. METHODOLOGICAL ISSUES AND MARGINAL COST ESTIMATES PER MODE OF TRANSPORT

The purpose of this section is to summarise our knowledge, as it is, about cost measurement in general and the computation of marginal costs of infrastructure use in particular. Results from four studies, which were

Table 1. Overview on Road Cost Allocation Methods Used in Different European Studies.

Country	Method Used
Austria	Regression analysis Adaptation of the German method
Denmark	Differentiation of capital and running costs into: Fixed costs, Vehicle-km (vkm)-dependent costs Space-dependent costs Weight-dependent costs Use of specific weight and space factors by type of vehicle
France	Differentiation between fixed and variable expenditures Use of different allocation factors such as: vkm weight-vkm standard axle-vkm
Germany	Differentiation between marginal costs and capacity costs Allocation of: Marginal costs by AASHO-road factors*vkm Capacity costs by (speed-dependent) equivalent factors*vkm
Italy	Differentiation between marginal and capacity expenditures, allocated by: vkm axle-weight-km standard-axle-km (Sakm)
Netherlands	Differentiation of investment expenditures and running expenditures into different sub-categories allocated by: vkm PCU-km axle load-km
Finland	Differentiation between fixed and variable expenditures allocated by: vkm weight-factors
Sweden	Differentiation of fixed and variable expenditures into vkm-dependent expenditures space- and speed-dependent expenditures (allocated by PCU-km) weight-dependent expenditures (allocated by AASHO-factor-km)
Switzerland	Allocation of: Weight-dependent costs of new investment (estimated by percentages per road type) by weight-factors Weight-dependent costs for pavement and investive maintenance by axle-load-vkm Capacity costs: 80% by vehicle-length*vkm 20% by vkm Current costs by vkm

Table 1. (*Continued*)

Country	Method Used
United Kingdom	Allocation of: Capital expenditure: 15% by max. GVW-km, 85% by PCU-km. Maintenance expenditure further differentiated by types of expenditures and different allocation factors applied Policing and traffic wardens by vkm.

Source: Link et al. (1999).

performed within the framework of the UNITE project and partly updated provide the core of the information. We start with the road sector (Section 3.1), where after rail (Section 3.2) is reviewed.

3.1. Road

The two sources to be discussed in more detail in this chapter are Link (2004) and Lindberg (2002). The first one describes an econometric study for Germany while the second applies an analytical approach for analysing marginal renewal costs for Swedish roads, also making use of engineering data.

Other recent econometric studies are Herry and Sedlacek (2002) and Schreyer et al. (2002). However, due to data limitations, these two studies were restricted to simple log-linear and linear regression models and faced serious problems with multicollinearity, difficulties to estimate models which reflect the joint impact of different vehicle types on road deterioration, etc. We will therefore not review them in detail here. Apart from these studies there exists almost no research that explores cost drivers for using existing road infrastructure.[2]

3.1.1. An Econometric Approach

The analysis described in Link (2004) makes use of a cross-sectional database, which contains a description of physical renewal measures on the German motorway network (e.g. length and type of renewal measure, material and construction type used, thickness of layers concerned) per motorway section (Table 2). This information is transferred into cost data by evaluating each measure with factor input volumes and factor input prices, and cumulated over a 20 year's time horizon in order to achieve a

Table 2. Parameter Estimates for Full Translog Models for German Motorway Renewal Costs[a].

	Model I (Eqs. (3.3)–(3.6))				Model II (Eqs. (3.7) and (3.4)–(3.6))			
	Coefficients	Std. dev.	t-value	Significance level	Coefficients	Std. dev.	t-value	Significance level
Constant	2.057	0.0253	81.174	0.000	−1.996	0.0547	−3.652	0.000
β_l	0.0028	0.0005	6.008	0.000	0.0026	0.0002	17.266	0.000
β_c	0.1530	0.0114	13.429	0.000	0.6428	0.0112	57.392	0.000
β_m	0.8442	0.0113	74.476	0.000	0.3546	0.0112	31.732	0.000
β_y	0.8754	0.0253	48.383	0.000	—	—	—	—
β_f	—	—	—	—	0.1457	0.0553	2.637	0.008
β_{fp}	—	—	—	—	−0.1219	0.0517	−2.357	0.018
β_{ll}	0.0015	0.0001	10.349	0.000	0.0015	0.0001	11.358	0.000
β_{cc}	0.1088	0.0059	18.186	0.000	0.1228	0.0053	23.238	0.000
β_{mm}	0.1105	0.0059	18.660	0.000	0.1244	0.0053	23.638	0.000
β_{yy}	0.1193	0.0066	17.935	0.000	—	—	—	—
β_{ff}	—	—	—	—	0.3752	0.1002	3.745	0.000
β_{pp}	—	—	—	—	0.1434	0.1289	1.112	0.266
β_{lc}	0.0001	0.0002	0.803	0.422	0.0000	0.0001	0.193	0.847
β_{lm}	−0.0016	0.0002	−9.652	0.000	−0.0015	0.0002	−9.586	0.000
β_{mc}	−0.1088	0.0059	−18.300	0.000	−0.1229	0.0053	−23.301	0.000
β_{yl}	−0.0001	0.0002	−0.423	0.672	—	—	—	—
β_{yc}	−0.1099	0.0051	−21.453	0.000	—	—	—	—
β_{ym}	0.1099	0.0051	21.577	0.000	—	—	—	—
β_{pf}	—	—	—	—	−0.2644	0.1070	−2.471	0.014

Source: Link (2004).

[a]The model contained also a vector of district dummies and a categorical variable indicating the type of material used for the upper layer. These variables are not reported here.

more densely populated data matrix and to smooth the cyclical pattern of renewal measures.

The study presents results from two translog cost functions estimated on these data. The first model analyses the economic process of renewing motorways as such. The renewal costs C_i are to be explained by the output variable Y_i defined as m^2 of motorways renewed at each section i over the period of analysis, a vector of input prices $W' = (w_{li}, w_{mi}, w_{ci})$ for labour, material and capital, and a set of dummy variables D_{ij} indicating the federal state j ($j = 1,\ldots,m$; $m = 8$) and M_{ik} representing the material used for renewals ($k = 1,\ldots, K$; $K = 7$). This basic model has the form

$$\begin{aligned}\ln C_i = {} & c + \sum_{j=1}^{m} \alpha_j D_{ij} + \sum_{k=1}^{K} \delta_k M_{ik} + \beta_y \ln Y_i + \beta_l \ln W_{li} + \beta_m \ln W_{mi} \\ & + \beta_c \ln W_{ci} + 1/2(\beta_{yy} \ln Y_i \ln Y_i + \beta_{ll} \ln W_{li} \ln W_{li} \\ & + \beta_{mm} \ln W_{mi} \ln W_{mi} + \beta_{cc} \ln W_{ci} \ln W_{ci}) + \beta_{yl} \ln Y_i \ln W_{li} \\ & + \beta_{ym} \ln Y_i \ln W_{mi} + \beta_{yc} \ln Y_i \ln W_{ci} + \beta_{lm} \ln W_{li} \ln W_{mi} \\ & + \beta_{lc} \ln W_{li} \ln W_{ci} + \beta_{mc} \ln W_{mi} \ln W_{ci} \end{aligned} \quad (3.3)$$

with the input cost share Eqs. (3.4)–(3.6) and the usual restrictions for homogeneity in input prices and for symmetry.

$$S_l = \frac{\partial \ln C_i}{\partial \ln W_{li}} = \beta_l + \beta_{ll} \ln W_{li} + \beta_{yl} \ln Y_i + \beta_{lm} \ln W_{mi} + \beta_{lc} \ln W_{ci} \quad (3.4)$$

$$S_c = \frac{\partial \ln C_i}{\partial \ln W_{ci}} = \beta_c + \beta_{cc} \ln W_{ci} + \beta_{yc} \ln Y_i + \beta_{lc} \ln W_{li} + \beta_{mc} \ln W_{mi} \quad (3.5)$$

$$S_m = \frac{\partial \ln C_i}{\partial \ln W_{mi}} = \beta_m + \beta_{mm} \ln W_{mi} + \beta_{ym} \ln Y_i + \beta_{lm} \ln W_{li} + \beta_{mc} \ln W_{ci} \quad (3.6)$$

A second model establishes the relationship between renewal costs and the traffic volume u_{fi} and u_{pi} of goods vehicles and passenger cars respectively, expressed as the annual average daily traffic (AADT) volume, and the afore mentioned dummy variables (cf. Eq. (3.7)). Again, the model includes the input cost share Eqs. (3.4)–(3.6) and imposes necessary conditions for homogeneity in input prices and for symmetry. Each of the two translog models described in Eqs. (3.3)–(3.6) and Eqs. (3.4)–(3.7), respectively provides a seemingly unrelated regression (SUR) model, which was estimated by means of the constrained ML estimator.

$$\ln C_i = c + \sum_{j=1}^{m} \alpha_j D_{ij} + \sum_{k=1}^{K} \delta_k M_{ik}$$
$$+ \beta_{\mathrm{l}} \ln p_{\mathrm{l}i} + \beta_{\mathrm{m}} \ln p_{\mathrm{m}i} + \beta_{\mathrm{c}} \ln p_{\mathrm{c}i} + \beta_f \ln u_{fi} + \beta_p \ln u_{pi}$$
$$+ \frac{1}{2}[\beta_{\mathrm{ll}} \ln^2 p_{\mathrm{l}i} + \beta_{\mathrm{mm}} \ln^2 p_{\mathrm{m}i} + \beta_{\mathrm{cc}} \ln^2 p_{\mathrm{c}i} + \beta_{ff} \ln^2 u_{fi}$$
$$+ \beta_{pp} \ln^2 u_{pi}] + \beta_{\mathrm{lc}} \ln p_{\mathrm{l}i} \ln p_{\mathrm{c}i} + \beta_{\mathrm{lm}} \ln p_{\mathrm{l}i} \ln p_{\mathrm{m}i}$$
$$+ \beta_{\mathrm{mc}} \ln p_{\mathrm{m}i} \ln p_{\mathrm{c}i} + \beta_{pf} \ln u_{fi} \ln u_{pi} \quad (3.7)$$

Almost all parameters of interest have the expected signs and are significant at the 5%, or at least the 10% level (Table 3). Exceptions are the interaction term between labour and capital, the second-order term for the traffic volume of passenger cars and some of the material dummies.

Estimating marginal renewal costs is rather straightforward. The cost elasticity, e.g. the ratio between marginal and average costs (MC/AC) can be derived as

$$\frac{\partial \ln C}{\partial \ln u_f} = \frac{\partial C}{\partial u_f}\frac{u_f}{C} = \beta_f + \beta_{ff} \ln u_f + \beta_{fp} \ln u_p \quad (3.8)$$

This elasticity was calculated at the mean value of u_p and ranges from 0.05 to 1.17, increasing with traffic but at a reducing rate (Table 3 and Fig. 1). By multiplying this elasticity with an average cost figure one can easily obtain a marginal cost estimate. Link (2004) reports marginal renewal costs ranging from 0.08€ to 1.87€/vkm. These rather high values are strongly influenced by the average cost per truck-km of 1.59€ which was calculated from the sample by assuming that all renewal costs are exclusively caused by trucks. The cost elasticity itself lies in a range which is comparable with other studies such as Lindberg (2002), which reports a range of 0.1–0.8 (Table 3) or Schreyer et al. (2002), which obtains a cost elasticity of 0.8.

3.1.2. An Analytical Approach

Costs for renewing the infrastructure, such as resurfacing roads or renewing tracks, are difficult to analyse based on empirical data. The main reason is that these measures typically are taken with long time intervals in-between, up to 30 or even 40 years for track renewal. While data on annual maintenance is scarce, time series of combined renewal and maintenance costs is virtually non-existent for such long time periods. Using observations for only a few years is not sufficient to establish how these renewal costs vary

Table 3. Marginal Renewal Cost Estimates and Cost Elasticities for Road Infrastructure.

	Germany	Sweden
Unit	€/vkm	€ cents/vkm
Marginal costs per truck-km	0.08 ... 1.87	0.77 ... 1.86
Cost elasticity (MC/AC) – range	0.05 ... 1.17	0.1 ... 0.8
Cost elasticity (MC/AC) – mean	0.87	0.4

Sources: Link (2004), Lindberg (2002).

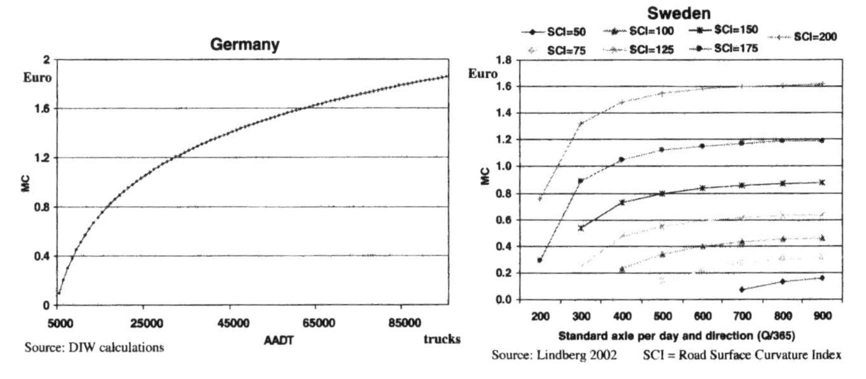

Fig. 1. Marginal Cost Curves for Motorway Renewals in Germany and Sweden.

with traffic, i.e. whether they are, in reality, part of the marginal costs for infrastructure use.

Lindberg (2002) therefore applies an analytical approach to estimate marginal costs for infrastructure renewal. In addition, engineering data are used to estimate critical parameter values. The application is used for road resurfacing, but the logic immediately carries over to any other mode or any periodic spending on renewal.

The logic of this approach is based on the observation that the length of an interval between two pavement renewals depends on the aggregate volume of traffic that has used a certain road section; standard axles is used as a measuring rod. Existing literature such as Newbery (1988b) seem to assume that the number of standard axles that can use a road before the pavement has to be renewed is a design parameter of road construction. Traffic load will therefore affect construction costs through the a priori design decision.

Lindberg's approach however makes use of the fact that the number of standard axles which the road can accommodate after all is a function of the *actual*, not the predicted traffic volume. The pavement will therefore be renewed when the road standard has deteriorated and reaches some pre-set number. Adding or subtracting vehicles to the original prediction will therefore affect the timing of a reinvestment and there is, consequently, a marginal cost associated with variations in traffic volume.

The model works as follows. The lifetime of a pavement (T) – the number of years between resealing – is a function of the (constant) annual number of standard axles that use the road (Q) and the strength of the road, where Θ denotes the number of 'standard axles' the pavement can accommodate, and m indicates the climate-dependent deterioration:

$$T = \left[\frac{\Theta(Q)}{Q}\right] \mathrm{e}^{-mT} \tag{3.9}$$

Each overlay has a cost of C. The first overlay takes place at year 0. We can then calculate the present value of an infinite number of overlays as (3.10) if considering the cost from the perspective of the initial overlay ($\mathrm{PV}C_0$) and as (3.11) if the perspective taken is that of a road which is t years old ($\mathrm{PV}C_t$); r is the relevant discount rate. To study the effect of annual traffic on the cost, the properties of the annualised present value of an average road ($\mathrm{ANC_{ave}}$) can be expressed as (3.12), where $\mathrm{PVC_{ave}}$ is just the average over roads of different ages. In this expression, the weathering effect is ignored.

$$\mathrm{PVC}_0 = C(1 + \mathrm{e}^{-rT} + \mathrm{e}^{-r2T} \cdots + e^{-rnT}), \qquad \lim_{n\to\infty} \mathrm{PVC}_0 = \frac{C}{(1 - \mathrm{e}^{-rT})} \tag{3.10}$$

$$\mathrm{PVC}_t = \mathrm{e}^{-r(T-t)} \frac{C}{(1 - \mathrm{e}^{-rT})} \tag{3.11}$$

$$\mathrm{ANC_{ave}} = r\mathrm{PVC_{ave}} = C/T \tag{3.12}$$

The marginal cost caused by shortening the renewal intervals due to higher traffic loads can be obtained by differentiating the annualised present value of the road with the annual traffic volume. By using the deterioration elasticity ε – the change of lifetime due to higher traffic loads (Eq. (3.13)) – and the definition of average costs as $\mathrm{AC} = C/T = C/QT$, the marginal costs for an average road MC_{ave} are related to the overall average cost as in (3.14).

$$\varepsilon = \frac{\mathrm{d}T}{\mathrm{d}Q}\frac{Q}{T} \tag{3.13}$$

$$MC_{ave} = -\varepsilon \mathrm{AC} \tag{3.14}$$

The decisive parameter for the relationship between the average and marginal cost is the value of the deterioration elasticity. The so-called fundamental theorem (Newbery, 1988a, b; 1989) says that average cost is equal to marginal costs. However, Eq. (3.14) illustrates that this is only valid if (a) there is no weathering effect and (b) if the number of standard axles the surface can withstand is constant, e.g. if the elasticity ε becomes negative unity. So the question then becomes empirical; can the assumption of zero weathering effect (or, more precisely, that there is no interaction between variations in traffic level and weathering) be substantiated and what is the value of ε?

Lindberg (2002) makes use of data from the Swedish Long Term Pavement Performance (LTPP) project, which has studied pavement deterioration since 1985. Today the programme consists of 639 road sections on 64 different roads. Most of the roads are located in the south and the middle of Sweden. Every year, a detailed distress survey, measurements of rut depth and the longitudinal profile is carried out at each section. The collected database also contains information about structural strength, surface condition, pavement structure, climate and traffic. An index which consists of three elements, namely the cracked surface, the longitudinal cracking and the transverse cracking is used as the quality statistic.

Based on this information, evidence that ε varies from -0.1 on high-quality roads with low traffic load up to -0.8 on low-quality roads with a high traffic load is given. For a certain traffic load, the (absolute value of) elasticity increases with reduced strength of the road and for a given strength the (absolute value of) elasticity increases with increased traffic. This result is intuitive appealing since the highest marginal cost can be expected at low quality roads with high traffic volumes and the lowest cost at high quality roads with low traffic volumes.

Two types of results with relevance for marginal cost estimation are provided by this analysis. In a first type of calculation, an average pavement cost and the deterioration elasticity were used to calculate marginal costs per standard axle on roads with different strength. The marginal costs lie in a range of 0.07€/100 Sakm up to 1.62€/100 Sakm (see Fig. 1).

Table 4. Average and Marginal Cost for the Swedish Road Sub-sample.

	Mean	Std. Dev.	Minimum	Maximum	Number of Cases
SCI[a]	133.997	44.3632	55.5224	269.104	249
Vehicles (AADT)[b]	5131.57	2278	1290	10900	249
Width (*m*)	11.7209	3.75126	7.5	20	249
Q (per day and direction)[c]	578.94	379.485	137	1320	249
Overlay cost (kEUR/km)	37.0	8.7	30.5	66.0	249
Lifetime (year)	11.8103	3.11661	3.36859	16.9688	249
Deterioration elasticity	−0.431342	0.221295	−0.80211	−0.00908	249
Average costs (EUR/Sakm)	0.022	0.016	0.006	0.093	249
Marginal costs (EUR/Sakm)	0.008	0.0061	0.0002	0.038	249

Source: Lindberg (2002).
[a]Road surface curvature index.
[b]AADT = Annual average daily traffic.
[c]Number of standard axles.

In a second type of calculation, information for a sub-sample of 249 road sections was used. For these sections, an average lifetime of 11.8 years and an elasticity of −0.43 was estimated. The marginal costs/100 Sakm were estimated to be 0.8€, assuming an average overlay cost of 2.2€/100 Sakm. The estimated costs per standard axle (see Table 4) are expressed as costs per vkm by using data from the Swedish Road Administration on standard axles per vehicle type for four groups of vehicles. According to this calculation, a marginal cost of 0.32€/100 vkm for light duty vehicles (LDV) and of 1.86€/100 vkm for the heaviest vehicles (HGV with trailer) was derived.

To interpret the results, one has to consider three basic assumptions of the analysis: (i) Climate conditions have no influence on the renewal interval. (ii) The age of roads is equally distributed within the whole road network. (iii) Pavement is renewed if the cracking index has reached a certain defined terminal value.

3.1.3. Summary

Both the econometric analysis for Germany and the engineering-based study for Sweden derive a gradually increasing marginal cost curve. These results are in contrast to the a priori expectation that in particular maintenance and renewal costs are proportionate to standard axles as it is suggested by the AASHO road test. But the econometric model that has been reviewed was based on observed spending, which does not necessarily correspond directly with road damages as measured within the AASHO road test. Neither can the approach chosen in Lindberg (2002) be compared directly with the design of the AASHO experiments which put road damages in direct relation to axle loads. In contrast to the AASHO test, Lindberg (2002) analyses how traffic load shortens the life-expectancy interval of a road overlay up to a threshold where it has to be renewed.

Moreover, the value of estimated marginal costs will be location-specific according to factor prices, traffic load, weather and ground conditions. Furthermore, both the approaches chosen and the cost categories, road types and vehicle categories included in the modelling differ. While both studies analyse the same type of cost (renewal costs) they differ regarding the approach (econometrics versus engineering-based approach), the network taken into account (motorways in Germany versus all roads in Sweden with a considerably lower share of motorways) and traffic density. A direct comparison of the marginal cost estimates from the two studies is therefore not recommended, neither is a transfer of marginal cost figures from one country to another.

For the purpose of comparison and generalisation of results, the cost elasticity should be preferred, since the elasticity measure avoids several of these problems. This elasticity ranges between 0.07 and 1.17 with a mean value of 0.87 for the German renewal cost analysis and is between 0.1 and 0.8 for the Swedish study.[3] However, also these figures should be used with caution.

A final observation refers to data availability and quality. The available experience has shown that sufficiently disaggregated, cross-sectional data both for maintenance and renewal costs as well as axle-load data and information on further influence factors is hard to obtain. For the econometric approach, the ability to estimate a full cost function with input cost share equations is often hampered by lack of cross-sectional information on input prices. On the other hand, in most countries the necessary long-term measurements on road conditions required for the approach chosen in Lindberg (2002) are not available, but could improve in future with the development of long-term pavement management systems. More than anything, the presentation therefore points to the need to perform more research, an issue that will be further emphasised in our concluding section.

3.2. Rail

Vertical separation of infrastructure from railway operations has become *a la mode* across the continent. An implication of the new industrial structure is that all costs related to infrastructure are now accounted for separately from train operation costs. This makes it obvious that the analysis of road sector marginal costs today has a direct parallel in the railway sector.

Our presentation of cost analyses of track infrastructure activities comprises three main parts. The first two provide reviews of studies undertaken as part of the UNITE project. Johansson and Nilsson (2004) is an econometric approach using Swedish data and Nash and Matthews (2002) reports about a British approach to estimating marginal costs, based on an engineering model. Third, an attempt is made to summarise the conclusions from these presentations and to contrast results to a third paper analysing French data, Gaudry and Quinet (2003).[4]

The study by Johansson and Nilsson (2004) also includes an analysis of Finnish data undertaken by these authors. RHK (2004) is a separate analysis performed by the Finnish Rail Administration on a somewhat larger data set that also includes spending on reinvestment. In view of the short time period and the major variations in reinvestment spending between years, it is no surprise that the explanatory power of the model falls. As previously discussed, reinvestment spending is difficult to analyse within the same framework as that used for understanding recurrent spending patterns, in particular for shorter periods of time. The results with respect to spending on reinvestment are therefore not reported here. Another analysis of marginal costs of track wear from Norway (Börnes–Daljord, 2003) makes use of similar techniques as those reported here, and are therefore not reviewed.

3.2.1. The Swedish Study

Sweden's 1988 vertical separation of its nationalised railways made *Banverket*, a government agency, responsible for the infrastructure, while the incumbent continued to operate all railway services. Since then, private operators have also started businesses and in June 2000, more than 20 firms operate railway services; cf. Nilsson (2002) for more institutional details.

The Swedish railway network (about 13,000 km) is split up into some 260 track units. Fig. 2 provides intuition about the structure of the micro-data used. In principle, a track unit is homogenous with respect to traffic and technical qualities. The single-track line between stations A and B is used by a certain number of trains and should therefore comprise one track unit. Between station B and a switch called C, the line is double tracked, while it is single tracked between C and D; these two parts of the line should be registered as separate track units. Since the traffic load differs, sections D–F, F–G and F–H should also comprise separate units. Major marshalling yards, such as E, are accounted for separately as are some major stations such as D.

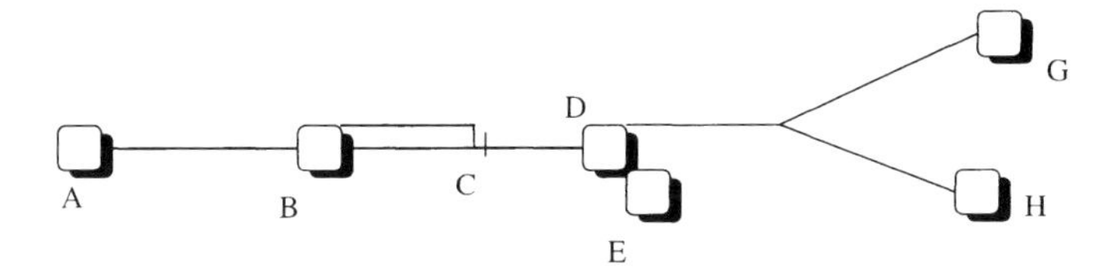

Fig. 2. Sketch of a Railway Network and its Track Units.

Except for spending on maintenance and traffic volumes, *Banverket* registers the number of switches, bridges, tunnels and other technical installations on each separate track unit. In addition, a track quality index comprising eight classes, each representing the (average) standard level of sleepers, fastenings and rails is assigned to each track unit. For instance, index no. 8 – the best tracks – is given to track sections with the predominantly heaviest (per metre) whole welded rails on concrete sleepers. An administrative distinction is also made between track units that are part of a main or secondary line and those that are not. 20 maintenance districts – each responsible for a set of track units corresponding to a certain number of track km – are responsible for recording the data. The dataset comprises the 1994–1996 period.

While common costs could in principle be distributed over the track units using different keys, the analysis is only applied to track-specific costs. The presumption is that the agency's common costs for main, regional or district offices do not vary with the traffic load; common costs affect average but not marginal, costs. Thus, the records account for about 1.6 out of the total spending on maintenance of 2.3 billion SEK in 1994. One implication is of course that this per se means that 30% of total costs are not to be made part of a marginal cost-charging tariff.

Traffic is measured as gross tonne km using each track unit, i.e. the number of km that trains of a certain weight – also including the weight of loco and cars – have traversed on each track unit. There are 169, 176 and 175 observations of costs per track unit available for the years 1994, 1995 and 1996.

The translog cost function of Berndt and Christensen (1972), presented in Section 2.1 above, was used as a flexible specification of the cost structure. The final model used for the analysis is given by

$$\ln C_{ijt} = \alpha + \alpha_t + \alpha_j + \beta_y^* y_{ijt} + \beta_u^* u_{ijt} + \beta_{yy} y_{ijt}^2 + \beta_{uu} u_{ijt}^2 + \beta_{yu} y_{ijt} u_{ijt} + \beta_I I_{ijt} + \beta_z' z_{ijt} + \varepsilon_t \quad (3.15)$$

Here, i is an index for track unit, j an index for district, t a time index and C_{ijt} the maintenance cost to be explained. Y_{ijt} is the track length and $y_{ijt} = \ln Y_{ijt}$, U_{ijt} is the utilization level (the gross weight of trains) and $u_{ijt} = \ln U_{ijt}$; z_{ijt} is a vector of track-technical variables (the number of switches, number of tunnels etc.); and ε_{jt} are assumed to be independently and identically distributed error terms. $I_{ijt} = 1$ if, at time t, track unit i of district j is located on a secondary/non-electrified line and zero otherwise.

Table 5 summarises the parameter estimates. Considering that cross-section data are used, the fit is excellent with $R^2 = 77\%$. The signs of the parameters of interest are, by and large, as expected. The parameters of main interest, track length (y) and utilization (u), are significant except for the second-order term for track length, significance or insignificance results referring to the 5% level at least.

Figs. 3 and 4 plot the elasticity of costs relative to line length and utilization (i.e. $\eta^{y}_{ijt} = \partial \ln C_{ijt}/\mathrm{d}y_{ijt}$ and $\eta^{u}_{ijt} = \partial \ln C_{ijt}/\mathrm{d}u_{ijt}$) against gross tonne km, Gtkm $= UY^{l}$, where Y^{l} is line length (line length, Y^{l}, differs from track length (Y) for track units with double tracks; $Y > Y^{l}$); each data point is simply $\eta^{u^*}_{ijt} = \beta^{*}_{ut} + 2\beta_{uu}u_{ijt} + \beta_{yu}y_{ijt}$ and correspondingly for line length. The elasticity of costs relative to utilization (η^{u}) falls when traffic loads increase and remain basically constant after the first two billion gross tonne km. The point-observations of elasticity therefore provide the image that the cost for maintaining tracks have the familiar u-shape, at least the falling part of the u. The same pattern prevails for η^{y}.

The mean values of these elasticities are summarised in Table 6. The overall mean of cost elasticity with respect to track length (η^{y}) is 0.80 while separate estimates for main and secondary lines are 0.71 and 0.97, respectively. Calculated standard deviations are, however, so large that we cannot safely claim mean elasticities to be below one.

However, for the main lines it is fair to say that increasing traffic load would reduce average maintenance costs. Our single most important observation is that the mean η^{u} is 0.17, meaning that the average costs for maintaining railway infrastructure decrease with the traffic load. This estimate will be used to calculate marginal costs.

One reason for differences between elasticities in the two classes of lines may be that main lines include track sections that are entirely or partly double tracked, while secondary lines are typically single tracked with fewer meeting stations. Since maintenance activities on main tracks can more often be undertaken with one of the tracks closed off from traffic for a relatively convenient period of time, the physically identical measures may require workers to get off the line more frequently on secondary lines; each

Table 5. Parameter Estimates from One Model Estimated on Swedish Data.

Variables/Coefficients	Estimate	t-Value
α	−6.75	−3.92
α_{95}	−0.005	−0.09
α_{96}	0.01	0.24
I/β_I	0.03	0.32
y/β_y^*	2.34	5.94
u/β_u^*	0.99	5.05
yu/β_{yu}	−0.1	−5.87
y^2/β_{yy}	−0.01	−0.29
u^2/β_{uu}	−0.01	−2.29
Bridge	0.005	0.708
Switches	0.01	3.60
Index	0.21	2.29
R^2	0.767	

All variable/coefficients are not reported in this table; cf. further Table 3 in Johansson and Nilsson (2004).

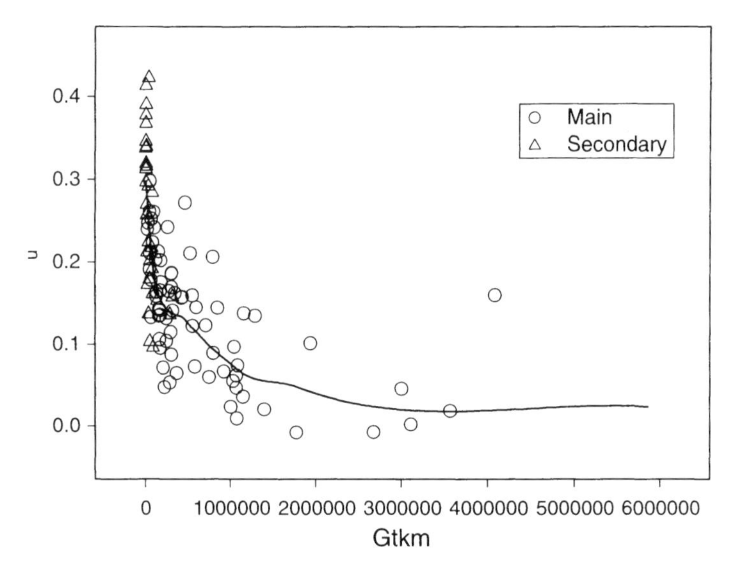

Fig. 3. The Elasticity of Utilisation on Cost against (Thousand) Gross tonne km. One Mark Represents 5 Data Points. The Line is Calculated Using a Loess Smooth.

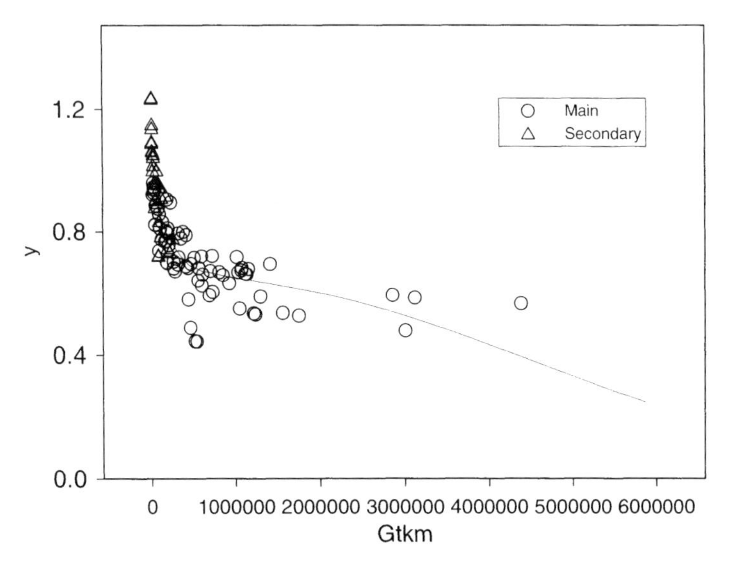

Fig. 4. The Elasticity of Track Length against (Thousand) Gross tonne km. One Mark Represents 5 Data Points. The Line is Calculated Using a Loess Smooth.

Table 6. Mean Elasticities of Track Length and Utilisation, Subdivided into Main and Secondary Lines.

	All	Main	Secondary
Mean η^{v}	0.80	0.71	0.97
Std. dev.	0.23	0.23	0.24
Mean η^{u}	0.17	0.14	0.23
Std. dev.	0.03	0.04	0.03

job thus takes more time. A second reason is that secondary lines more often are of secondary standard, i.e. with tracks that are not whole-welded, etc. Older equipment is more costly to maintain.

The elasticities reported in Table 6 are used to derive estimates of marginal costs for each single track unit, i.e. of the wear and tear inflicted on tracks by allowing for an additional train of some certain weight to pass over a track unit. To do so, it is necessary to include a distance component in the traffic activity measure. While elasticity is measured relative to an increase in gross tonne per track unit, the preferred marginal cost measure is due to an increase in Gtkm. The measure of marginal cost (MC) per Gtkm is then

$$\mathrm{MC} = \frac{\partial C}{\partial \mathrm{Gkm}} = \frac{\partial \ln C}{\partial \ln \mathrm{Gkm}} \frac{C}{\mathrm{Gkm}} = \frac{\partial \ln C}{\partial \ln U} \frac{C}{\mathrm{Gkm}} = \eta_u \frac{C}{\mathrm{Gkm}}$$

It is thus assumed that, at the margin, the cost is unaffected by line length Y^l and hence MC is the marginal cost of increasing the utilization for a given line length. Estimates of the marginal cost are given by substitution of the estimates and the inclusion of fitted costs.

To provide a comprehensive statistic, the 'average marginal cost' has been calculated, both for the network as a whole and for the main and secondary lines separately. For this purpose, the respective marginal costs are weighted together using the traffic activity on each track unit as weight, i.e. $weight_{ijt} = \mathrm{Gtkm}_{ijt}/\sum_{ij} \mathrm{Gtkm}_{ijt}$. This weighting scheme generates the same level of revenue as if a separate charge is levied for each track unit. The estimated marginal costs are shown in Table 7, both for the network as a whole and separated for main/electrified and secondary/non-electrified lines.

An implication of these numbers is that a 1300 gross tonne freight train, moving some 800 km between its points of departure and arrival and being charged (on average) 0.0012 SEK per gross tonne km would have to pay about SEK 1200 for the whole trip. Since we have only included 70% of the annual spending on track maintenance in Sweden, about 12% of the total costs would be recovered.

A final word about the basis for the observations – i.e. the underlying maintenance policies of the respective agencies – is warranted. Results provide an impression of how costs are affected by different explanatory variables, in particular traffic. The basic result is that maintenance activities are not very responsive to variations in traffic. There is, however, at least some causal relationship between traffic and maintenance. One reason might be that agencies have a policy of fixing problems when these pop up.

Table 7. Estimates of Marginal Cost in Öre per Gross Tonne km (1SKR = 100 öre).

	1995	2000
All	0.117	0.120
Primary railways	0.082	0.084
Secondary railways	0.909	0.930

A positive relation between traffic and costs then simply indicates that more must be done the more extensive is the traffic, the reason being that more traffic will wear down the infrastructure more quickly. Another, or rather a complementary, policy could be that tracks are to be held in reasonable shape, irrespective of problems that actually occur. This policy could, in turn, be based on traffic loads, which would then explain the positive relationship. It is, however, not possible to know which explanation is most reasonable for describing the results.

3.2.2. The British Study

As part of the periodic review of infrastructure charges in the year 2000, the British rail infrastructure manager, Railtrack, developed a bottom-up model to estimate marginal usage costs. This starts from detailed engineering relationships and adds up individual elements of cost caused by additional trains where they can be identified. The resulting sum of individual cost elements has then been calibrated in order to fit with the total costs derived from the organisation's budget allocations for maintenance purposes.

In addition, the Rail Regulator has used a consultant to develop an alternative top-down model. This is also based on engineering relationships developed by Railtrack. It starts by identifying the total planned maintenance and renewal expenditure on different types of assets, namely track, signalling, electrification equipment and structures. It then estimates the percentage of these costs which vary according to the number of trains run so as to derive a total variable cost for each asset type. Finally, detailed engineering relationships are used to allocate these total variable costs to particular vehicle types.

Research carried out for Railtrack indicated that ~20% of its total maintenance and renewal costs vary with asset usage. The variable cost component is dominated by track maintenance and renewal, with 78% of costs varying with usage; for structures maintenance the figure is 12% and for signalling and electrification equipment 5%. Assets such as parking and housing, rail stations, etc. were excluded from the analysis.

While Railtrack believe their estimates of the extent of cost variability is broadly in line with experience from elsewhere, two caveats should be noted. First, the British rail network, particularly in comparison with the U.S., has a much greater share of its total costs in signalling and electrification, which are inherently less variable with volume than is track. Second, traffic densities in Britain are relatively low in international terms. It would therefore be expected that the proportion of variable costs would be much lower for Britain than, for example, for the U.S.

The Rail Regulator's consultants furthermore suggested that the percentage of variable track maintenance and renewal costs actually differs depending on track standard and traffic volume. For example, for lines with low traffic load the fixed costs will dominate the variable costs such that they will experience proportionately little change in total cost, and vice versa for lines with high traffic volumes. Other things being equal, the incremental cost will generally be greater, the higher the track quality and the higher the average line speeds, although this will also be a function of the standard of initial construction. Other research has furthermore linked track maintenance and renewal costs to speed, axle loads and the condition of the track. It has established that costs vary at ~60–65% of the rate of change in both speed and axle load. The rate of change was also sensitive to track condition, with the increase generally being greater the poorer the quality of the track.

Table 8 summarises the estimates made by the consultants of how different cost components vary with usage for the British rail network. The figures are based on both domestic British experience and on reviews of international studies. In this, it relies heavily on judgement. Based on this top-down approach to estimate the overall variable costs, estimates are made to apportion the costs per vehicles. Some examples of the resulting figures are given in Table 9.

The described approach estimates AVC and assumes that this may be taken to approximate marginal cost (MC). It is obvious that AVC is equal to MC when AVC is constant. It seems reasonable to assume that the damage done by a particular run of a vehicle over a stretch of track will not depend on the density of traffic on that track in terms of trains per week. It may, however, depend on the cumulative volume of traffic since the track was last repaired or replaced. But this should not provide reason to vary charges, given the fact that each train will run over a variety of segments of track each at a different stage in their life cycle.

3.2.3. Summary

What do we learn about the structure of track maintenance costs from the different studies reviewed above? With respect to the econometric approach it is apparent that the translog specification used on Swedish data to model the functional relationship between costs and explanatory variables seems to provide a good basis for understanding the pattern of spending on track maintenance. Costs vary non-linearly with variations in traffic and track length, but non-linearities are not very strong.

Table 8. Percentage of Infrastructure Cost that Varies with Use.

	Percent Variable	Percent Variable by Asset Category
Track		38
Maintenance	30	
Renewals		
Rail	95	
Sleepers	25	
Ballast	30	
S&C	80	
Structures	10	10
Signals		2
Maintenance	5	
Renewals	?	
Electrification		24
Maintenance		
AC	10	
DC	10	
Renewals		
AC	35	
DC	41	

From Nash and Matthews (2002, Table 3).

Table 9. Examples of Usage Charges, Pence per vkm (1999/2000).

Diesel Shunter (Class 08)	2.6
Diesel loco (class 47)	63.9
Electric loco (class 90)	59.7
Passenger car (mk 3)	10.4
Diesel multiple unit (class 158)	10.4
Electric multiple unit (class 333)	
Powered car	15.4
Trailer car	11.9
Freight wagon (pence/Gtkm)	2.7–3.3[a]

[a]From Nash and Matthews (2002).

A second observation is that the explanatory power of the model is high; information about traffic levels, track length, number of switches and quality provides a good intuition for understanding the levels of spending on track maintenance. When information about these variables is available, the classification of tracks to one administrative class or another does not add to our understanding of maintenance patterns.

Third, the research demonstrates that the costs for maintaining tracks constitute a major reason for the scale economies of the industry, at least for the levels of track use reported here; the cost elasticity for marginal variations in traffic levels is far below one. One policy implication is that if traffic is priced at marginal costs for track wear, revenue from these charges would be inadequate for recovering the total costs for the maintenance of the railway infrastructure.[5]

The technique to estimate cost elasticities as a step towards calculating marginal costs also has a merit in that it facilitates cross-country comparisons. While the *level* of cost may vary according to country-specific idiosyncrasies, the elasticity measure only expresses relative numbers and may make it easier to compare the situation in one country with that in another.

The British approach, fourth, also suggests a line of reasoning which may be of some relevance to other countries. Suppose that there are two different railways with cost breakdowns as summarised in Table 10. If we think about the overall percentage of cost variability as an estimate of the cost elasticity for infrastructure costs, we may estimate the cost elasticity of railway *A* as 0.3 and that of railway *B* as 0.25. Given that the cost elasticity is the ratio of marginal to average costs, and if there is information about total infrastructure costs for each railway, it is feasible to estimate marginal costs. It might also be feasible to use the relativities from the British study to estimate the marginal costs for different types of rolling stock, if that is needed.

The potential problems generated by both the econometric model used on Swedish data and the engineering approach used in Britain, are however eloquently illustrated by a recent study of French data. Gaudry and Quinet (2003) have access to a database covering over 1,500 sections of the French

Table 10. Example of a Generic Model for Calculating Marginal Costs.

	Share of Total Cost		
	Railway *A*	Railway *B*	Variability (%)
Track	70	50	38
Structures	10	10	10
Signals	10	20	2
Electrification	10	20	24
Overall elasticity	0.3	0.25	

From Nash and Matthews (2002).

rail infrastructure in 1999, representing about 90% of the total 40,000 km network. Costs for infrastructure maintenance and operations are reported for each of these sections. Technical information relates to length of each section and number of tracks, the type of power supply, the type of safety control, the type and age of sleepers, the number of track apparatus, the maximum allowed speed (which works as a quality indicator) and whether the track is a high-speed line or not. The traffic information describes the number of train services for each type of train and the total weight of the traffic for these services; train-types comprise long-distance passenger, regional passenger, Paris suburban, freight and "other" services.

Based on this huge data-base, the authors are able to evaluate the use of different econometric specifications of the model, the Translog reported about being just one. It turns out that this way to specify the equation is *not* to be preferred. The following list points to some other results of this specification.

(1) There seems to be reason to account for output in a more disaggregate way, in particular to use number of trains and (average) train weight rather than the total number of tons.
(2) There seems to be an axle load damage function that differs across train services; increasing average weight of passenger trains increases costs proportionately, but increasing the average weight of freight trains increases costs much more than proportionately.
(3) The higher the maximum permissible speed, the more is spent on infrastructure investment and the lower is the maintenance cost.
(4) Marginal costs depend on the kind of service, meaning that a high-speed train does not have the same marginal cost as a freight train.
(5) The rate of cost recovery of a system with marginal cost pricing differ according to the level of traffic, with lines with average traffic recovering 37% of costs and lines with at least five times the average recovering more than the full costs.

Several of these results are at odds with other results reported previously in this chapter. It remains to be seen in a final version of the analysis of French data whether these results still are at large. Irrespective of which, the paper illustrates the importance of using a comprehensive database, analysed by skilled researchers. Many of the problems and question-marks that have been identified throughout our study emanate from the lack of data and the need to make assumptions in order to have at least something to say about the problem at hand.

4. CONCLUSIONS FOR POLICY AND RESEARCH

In our introduction, we pointed to two recurrent issues in the estimation of marginal infrastructure costs; the relationship between maintenance costs related to wear and to weather, respectively, and the appropriate way to handle reinvestments in the marginal cost context. The answer to the first question seems to be that weathering explains much of the spending on railway infrastructure, with use accounting for well below 50% of maintenance costs; our best-guess elasticity of costs relative to use lies between 0.15 and 0.30. The variation across road types and across light and heavy traffic seems to be higher for road infrastructure, with elasticities sometimes even being above 1. Since a substantial part of costs are accounted for at central-office levels and not included in the analysis, cost recovery is in optimum even lower than these numbers indicate.

Moreover, spending on reinvestment should be made part of the marginal cost for infrastructure use, but because of the long-time periods between each renewal, the same approach as for analysis of maintenance activities cannot be used. A possible approach is to cumulate long-term observations on renewal spending in order to smooth the cyclical pattern. An alternative, combined analytical and engineering approach has been suggested. Again, also parts of the reinvestment costs may have to be deleted from the tariff since it most probably is due to the weathering effect. If so, the deficit of a marginal cost pricing policy would be even larger.

We have arrived at these admittedly very sketchy conclusions using two different methods referred to as engineering and econometric approaches, respectively. The engineering approach benefits from the possibility to discuss the problem on a comprehensive basis, using a combination of experience and pragmatic rules of thumb. Econometric methods analyse recorded behaviour, i.e. historical data. With hindsight, these data may represent flawed policies, for instance, since decision makers had incomplete information when allocating resources or in view of external budget constraints that are seen to be inappropriate. Spending in 1 year may also reflect the accumulated consequences of traffic loads over previous years. With the data at hand, we have poor possibilities of controlling for any of these circumstances.

The strength of the econometric analysis is, on the other hand, that results – at least in their ideal versions – can be based on very few assumptions and that the researcher simply looks at the evidence at hand. This is in obvious contrast to engineering approaches which often have to make "reasonable" assumptions. For all the skills of experts, there is always a risk of being blind for patterns that would be discerned in systematic analyses of larger databases. There is also an obvious risk for falling prey to all sorts of biases, for instance, about what different experts believe to be "appropriate".

A common problem for the engineering and econometric approaches is the lack of sufficiently disaggregated data about the way in which infrastructure costs, conditions and use develop over time. To our mind, infrastructure agencies should in principle be collecting such data on a continuous basis. But in the cases when they actually have data, it is often difficult to get access to them, the proclaimed reason being confidentiality.

The attitude taken by the respective authorities towards the need to collect relevant data seems to be overly relaxed. The whole policy endorsed by the EC is based on that marginal costs can be calculated. In a complex reality, i.e. in a vast rail and road infrastructure which varies across countries in numerous ways, this estimation cannot be expected to be simple.

The basic data that we need to carry out the analysis is, on the other hand, not different from information required for running any reasonably well-managed business. We have demonstrated that what really is needed is information about costs for maintenance and reinvestment, about the traffic that makes use of the facilities as well as appropriate descriptive data of infrastructure qualities, all on a sufficiently disaggregate level. Most of this information is being collected for quite other reasons. What is needed is, therefore, primarily that the senior management takes this task seriously and actually updates the necessary data on an annual basis.

The results presented here primarily reflect a first go at analysing costs for providing road and railway infrastructure. We have strong reason to believe, not least having read the draft study of French data, that better information may change our conclusions radically. It is, therefore, not worthwhile to try to squeeze very much more information out of existing data; it is more appropriate to outline a 5-year plan for data collection where we know in beforehand what is needed. In spite of the results summarised in the present chapter, the jury is still out considering the verdict on many if not most policy issues that have been raised.

NOTES

1. Schreyer et al. (2002) had only data from one WIM station.
2. Talvitie and Sikow (1992) analyse the features of the construction process of new highway projects in Finland and focus on the productivity measurement of this

process. Since their cost function does not contain any traffic variable it is not possible to derive marginal costs of infrastructure use from their research.

3. Schreyer et al. (2002) report a cost elasticity between 0.72 and 0.82 derived with a log-linear regression model.

4. The motive for not making a detailed review of this paper is that it is yet only in a preliminary form.

5. Another component of the marginal cost is scarce track capacity. Nilsson (2002a) suggests a way of also charging for this cost. Charging for congestion costs would make up for at least parts of the deficit from charging wear and tear.

REFERENCES

Bauer, P. W. (1990). Recent developments in the econometric estimation of frontiers. *Journal of Econometrics*, *46*(1/2), 39–56.

Berndt, E. R., & Christensen, L. R. (1972). The translog function and the substitution of equipment, structures and labour in U.S. manufacturing, 1929–1968. *Journal of Econometrics*, *1*, 81–114.

Berndt, E. R., & Khaled, M. S. (1979). Parametric productivity measurement and choice among flexible functional forms. *Journal of Political Economy*, *87*, 1220–1245.

Booz, Allen, & Hamilton (1999). *Railway infrastructure cost causation*. Report to Office of the Rail Regulator, London.

Börnes–Daljord, Ö. (2003). *Marginalkostnader i jernbanenettet*. [Marginal costs in the railway network; in Norwegian only.] Rapport 3/2003 from the Ragnar Frisch Centre for Economic Research available at www.frisch.uio.no

Cantos, P. (2000). A subadditivity test for the cost function of the principal European railways. *Transport Reviews*, *20*(3), 275–290.

Caves, D. W., Christensen, L. R., Tretheway, M. W., & Windle, R. J. (1985). Network effects and the measurement of returns to scale and density for U.S. railroads. In: A. F. Daughety (Ed.), *Analytical studies in transport economics*. Cambridge: Cambridge University Press.

Christensen, L., Jorgenson, D., & Lau, L. (1975). Transcendental logarithmic utility functions. *American Economic Review*, *65*, 367–383.

Daughety, A. F., Nelson, F. D., & Vigdor, W. R. (1985). An econometric analysis of the cost and production structure of the trucking industry. In: A. F. Daughety (Ed.), *Analytical studies in transport economics*. Cambridge: Cambridge University Press.

Davidson, R., & MacKinnon, J. (1985). Testing linear and loglinear regressions against Box–Cox alternatives. *Canadian Journal of Economics*, *18*, 499–517.

De Borger, B. (1992). Estimating a multiple output generalised Box–Cox cost function. *European Economic Review*, *36*, 1379–1398.

Gaudry, M., & Quinet, E. (2003). *Rail track wear-and-tear costs by traffic class in France*. Montreal: Universite de Montreal, Publication AJD-66.

Grabowski, R., & Mehdian, S. (1990). Efficiency of the railroad industry: A frontier production function approach. *Quarterly Journal of Business and Economics*, *29*(2), 26–42.

Greene, W. H. (2000). *Econometric analysis*. New York: Prentice-Hall.

Herry, M., & Sedlacek, N. (2002). Road econometrics – Case study motorways Austria. UNITE (*UNIfication of accounts and marginal costs for Transport Efficiency*) *Deliverable 10*,

Annex A1c. Funded by EU 5th Framework RTD Programme. ITS, Leeds: University of Leeds. http://www.its.leeds.ac.uk/projects/unite/

Highway Research Board (1961). *The AASHO-road-test – history and description of project*. Special Report 61 A, Washington, DC.

Himanen, V., Idstrom, T., Göbel, A., & Link, H. (2002). Case study Helsinki–Vantaa Airport *UNITE (UNIfication of accounts and marginal costs for Transport Efficiency) Deliverable 10*, Annex A5. Funded by EU 5th Framework RTD Programme. ITS, Leeds: University of Leeds. http://www.its.leeds.ac.uk/projects/unite/

Johansson, P., & Nilsson, J. E. (2004). An economic analysis of track maintenance costs. *Transport Policy, 11*, 277–286.

Keeler, T. E. (1974). Railroad costs, returns to scale and excess capacity. *Review of Economics and Statistics, 56*, 201–208.

Lardner, D. (1850). *Railway economy: A treatise on the new art of transport. Dts management, prospects and relations*. New York: A.M. Kelley (1968 reprint).

Lindberg, G. (2002). Marginal costs of road maintenance for heavy goods vehicles on Swedish roads. *UNITE (UNIfication of accounts and marginal costs for Transport Efficiency) Deliverable 10*, Annex A2. Funded by EU 5th Framework RTD Programme. ITS, Leeds: University of Leeds. http://www.its.leeds.ac.uk/projects/unite/

Link, H. (2004). *An econometric analysis of motorway renewal costs in Germany*. Discussion Paper no. 9, Münster: University of Münster.

Link, H., Dodgson, J., Maibach, M., & Herry, M. (1999). *The costs of road infrastructure and congestion in Europe*. Heidelberg: Physica/Springer.

Lorenz, M. O. (1916). Cost and value of service in railroad ratemaking. *Quarterly Journal of Economics, 21*, 205–218.

Nash, C., & Matthews, B. (2002). British rail infrastructure case study. *UNITE (UNIfication of accounts and marginal costs for Transport Efficiency) Deliverable 10, Annex A4*. Funded by EU 5th Framework RTD Programme. ITS, Leeds: University of Leeds. http://www.its.leeds.ac.uk/projects/unite/

Newbery, D. M. (1988a). Road users charges in Britain. *The Economic Journal, 98*, 161–176.

Newbery, D. M. (1988b). Road damage externalities and road user charges. *Econometrica, 56*, 295–316.

Newbery, D. M. (1989). Cost recovery from optimally designed roads. *Economica, 56*, 165–185.

Nilsson, J-E. (2002). Restructuring Sweden's railways: The unintentional deregulation. *Swedish Economic Policy Review, 9*(2), 231–254.

Nilsson, J-E. (2002a). Towards a welfare enhancing process to manage railway infrastructure access. *Transportation Research, Part A, 36*, 419–436.

Oum, T. H., & Waters, W. G., II (1998). Recent developments in cost function research in transportation. In: G. De Rus & C. Nash (Eds), *Recent developments in transport economics* (pp. 33–73). Aldershot: Ashgate Publishing.

RHK (2004). Marginal rail infrastructure costs in Finland 1997–2002. *Report from the Finnish rail administration available at* www.rhk.fi.

Ripley, W. Z. (1923). *Railroads: Finance and organisation*. New York: Longmans, Green and Co.

Schreyer, C., Schmidt, N., & Maibach, M. (2002). Road econometrics – Case study motorways Switzerland. *UNITE (UNIfication of accounts and marginal costs for Transport Efficiency) Deliverable 10, Annex A1b*. Funded by EU 5th Framework RTD Programme. ITS, Leeds: University of Leeds. http://www.its.leeds.ac.uk/projects/unite/

Shephard, R. W. (1970). *The theory of cost and production*. Princeton: Princeton University Press.

Small, K. A., Winston, C., & Evans, C. A. (1989). *Road work: A new highway pricing and investment policy*. Washington: The Bookings Institution.

Talvitie, A. P., & Sikow, C. (1992). Analysis of productivity in highway construction using alternative average cost definitions. *Transportation Research*, *26B*(6), 461–478.

Wechsler, M. (1998). Analyse des Schwerverkehrs und Quantifizierung seiner Auswirkungen auf die Straßenbeanspruchung. *Straße und Autobahn*, *8*, 402–406.

OPERATING COSTS

Ofelia Betancor, Miguel Carmona, Rosário Macário and Chris Nash

1. INTRODUCTION

This chapter is concerned with measuring marginal transport supplier operating cost. By operating costs, we mean those costs incurred by operators of transport vehicles in providing transport services. We exclude taxes and payments for services such as the provision and maintenance of infrastructure or for accident or environmental costs as these cost elements are dealt with in other chapters.

As with infrastructure costs, three primary approaches for marginal cost estimation can be identified:

- the econometric approach, which uses statistical techniques such as multiple regression analysis to estimate the relationship between costs and outputs;
- the engineering approach, which uses engineering relationships to estimate the relationship between inputs and outputs and attaches prices to the inputs to obtain costs;
- the cost allocation approach, which takes accounting information, divides it into fixed and variable elements and allocates the variable elements to appropriate measures of output.

In practice, firms in the transport industry tend to use cost allocation methods, perhaps informed by engineering and/or econometric relationships. Researchers tend to use the econometric approach. All three will be

Measuring the Marginal Social Cost of Transport
Research in Transportation Economics, Volume 14, 85–124
Copyright © 2005 by Elsevier Ltd.
All rights of reproduction in any form reserved
ISSN: 0739-8859/doi:10.1016/S0739-8859(05)14004-9

discussed in what follows, and some comparisons made between them to try to understand the role of each approach.

In the next section, we consider some fundamental issues regarding operating cost measurement methodology. We then consider in turn air, rail and urban transport[1] before reaching our conclusions.

2. METHODOLOGY

When analysing the costs of providing a transport service, there are three main issues to consider:

- how to measure output?
- what categories of cost are relevant?

and

- what are the main cost drivers?

Measurement of output has been a contentious issue in the literature. Much cost estimation treats the output as passenger kilometres or freight tonne kilometres transported. For transport provided for a single specific customer, and where the volume is known in advance, this may be appropriate, although it will still be necessary to allow for the effect on costs of characteristics such as consignment size, length of haul and time of day/week/year. However, for scheduled transport services, this measure mixes up supply and demand. What the operator supplies is a given level of service; demand determines how well it is used. Thus, it is more appropriate when analysing costs to use a measure such as vehicle or train kilometres. However, if ultimately we are concerned with the estimation of the marginal costs of carrying additional traffic, it is necessary to translate cost estimates per vehicle kilometre into costs per passenger or freight tonne kilometre, using suitable assumptions about marginal load factors. This issue is returned to in Chapter 5 (this volume).

If a cost allocation approach is taken for marginal cost calculations, all supplier operating costs (SOCs) must be classified as variable costs or fixed costs. In theoretical terms, this distinction is very easy to establish: variable costs would be all categories of expenditure that vary according to the level of transport service provided, while fixed cost would have to be paid by the company even if the level of service were zero. Only variable costs must be considered when calculating marginal costs, which can be estimated by dividing variable costs by the level of service. The only problem with such a categorisation is that sooner or later all factors become variable, which obliges us to make a further distinction between the short and the long run period. For instance, consider the cost of personnel. In theory, this could be regarded as a variable cost, since if the level of service is zero, the company could dismiss all employees and reduce that cost to zero. However, part of the staff of a company is required regardless of the actual level of services (e.g. general administration), while some others could be more easily considered as variable (e.g. vehicle drivers). For marginal cost estimation, it is more useful to ask, how variable is labour for modest changes in service levels (e.g. $\pm 10\%$) rather than whether it could be avoided if services ceased altogether.

It is usually considered that operating costs may be related to three specific facets of transport operations:

(1) the number of vehicles required
(2) the hours the vehicles are in service and
(3) the distance they operate.

For a simple cost allocation approach, it is necessary to allocate all variable costs to one or other of these output measures. For instance, a typical approach would be to allocate as follows:

1. *Vehicle-related costs*: Includes annual depreciation of equipment, interest and leasing charges, maintenance and repairs. All costs related to essential personnel and supplies required for these activities.
2. *Time-related costs*: All personnel required for service to passengers and freight: crew, stewards, inspectors, catering staff.
3. *Distance related costs*: Fuel, oil, tyres.

An advantage of the econometric approach is immediately apparent, in that some of these cost categories (e.g. maintenance) may be a mixture of a fixed amount per vehicle plus time- and distance-related elements. An econometric approach would in principle be able to allow for the impact of all three variables on all categories of cost, as well as allowing for non linearities in the relationship. As against that, the econometric approach has big problems with data. For instance, information available will seldom allow for all three variables to be used as explanatory variables, and if it did they might be so collinear that it was impossible to disentangle their effects.

The reason for identifying separately costs associated with these different elements of output is that services differ with respect to the mix of number of vehicles required, distance and time spent running. All these impact on costs. For instance, a service which runs day long using just one vehicle and achieving high average speeds will have much lower costs per vehicle kilometre than a service which covers the same distance but at low speeds and with a peak requirement for several vehicles. Thus an option in econometric work is to try to introduce other variables (e.g. distance travelled by fast versus stopping services, mean service speed), which will pick up this effect.

As well as the character of the services operated, the character of the vehicles used (e.g. single versus double deck, seating capacity, floor area per passenger) will influence costs. In a cost allocation approach it will be necessary either to obtain breakdowns of each cost category by type of vehicle, or to estimate some relativities (e.g. one double deck bus equals 1.5 single deck, perhaps using engineering relationships such as fuel consumption for fuel costs to establish the relativities). Again, an econometric equation could in principle introduce separate variables for the number of vehicles of each type and for the hours and distance operated, but data problems and multi collinearity make this problematic. An alternative is to use measures of output such as seat kilometres or gross tonne kilometres, but these will typically cost more to produce using a large number of small vehicles than a small number of large, so a measure of vehicle size may be introduced as an additional variable.

As mentioned above, the other key cost driver is time of day. If an additional service is to be operated at a time when both vehicles and staff are sparse, then the marginal cost may be very low. On the other hand, if a short additional trip requires an additional vehicle and additional staff, then the marginal cost may be very high. The best way of establishing this is to run a vehicle and crew scheduling programme, which will identify the change in the number of vehicles and crew needed for any particular change in schedule. In the absence of this a simple approximation into peak and off-peak is the norm. It is usually assumed that vehicles are fully utilised in the peak but not in the off-peak, but the same may not necessarily apply to staff, where a judgement has to be made in each specific case.

In short, whether we are using cost allocation, engineering or econometric methods, we need to pay careful attention to the measurement of output, subdividing it into categories according to type of service, type of vehicle and time of day. The exact approach taken will need to be pragmatic according to the data available and the characteristics of the service in question. This will be illustrated by a variety of approaches taken in the case studies discussed in the following sections.

3. AIR TRANSPORT

3.1. Introduction

The categorisation of costs constitutes an essential task in an airline decision-making process. It facilitates new investment evaluation or the adoption of adequate pricing policies (Doganis, 1995). Classification of costs performed is the result of accounting practices, though it is also affected by the International Civil Aviation Organization (ICAO) regulations and recommended procedures. At ICAO, the normal practice is to distinguish, first of all, between *operating* and *non-operating* costs, so as to separate costs derived from the main activity of an airline from those cost items that are generated by operations not related with its main objective or that do not have a close relationship with the typical operation of air services. Financial costs, differences between equipment residual and market realisation values, losses arising at money exchange operations or losses from affiliated companies, are all instances of non-operating costs. Operating costs are further divided into *direct* and *indirect*. Table 1 illustrates the classification of airlines' cost.

The distinction between direct and indirect operating costs derives from the fact that some costs might change with type of aircraft (direct), while other cost items would remain unaffected (indirect). However, the division of operating costs into these categories is not always a clear task, and some carriers might be even applying contradictory accounting rules. ICAO considers three main types of direct costs (flight operations, maintenance and depreciation), and five other categories as indirect cost (user charges, passengers services, ticketing and sales, general/administrative and others).

Regarding *flight operation* costs, the first component included is the flight crew personnel cost (pilot, co-pilot and flight engineer). Other elements considered under this heading are: aircraft fuel and oil, flight equipment insurance, rental of flight equipment, flight crew training (not amortizable) and other flight expenses. The flight crew cost is usually expressed as an hourly rate per type of aircraft, whereas, given that fuel consumption varies not only with type of plane and route length, but also with meteorological conditions and flight altitude, fuel cost is commonly presented on a route basis. The oil consumption is calculated according to type and number of engines, and it is also expressed as an hourly rate.

The flight equipment insurance item can be calculated as a percentage of the purchase price. Insurance premiums vary mainly with the geographical

Table 1. ICAO Structure of Airline Costs.

Operating costs	Direct operating costs	Flight operations
		Flight crew salaries and expenses
		Aircraft fuel and oil
		Flight equipment insurance
		Rental of flight equipment
		Flight crew training (not amortizable)
		Other flight expenses
		Maintenance and overhaul
		Depreciation and amortization
		Flight equipment
		Ground property and equipment
		Others
	Indirect operating costs	User charges and station expenses
		Landing and associated airport charges
		En-route facility charges
		Station expenses
		Passenger services
		Ticketing, sales and promotion
		General and administrative
		Other operating expenses
Non-operating costs	Gains or losses derived from assets retirement	
	Net interest payments	
	Profits or losses from affiliated companies	
	Government subsidies or payments	
	Other non-operating items	

area where the airline operates, and may be converted into an hourly rate as well dividing it by estimated utilisation block hours per each type of plane.[2]

Flight equipment may be rented or purchased. If the carrier decides to rent it, the cost incurred will be incorporated as flight operation costs. It is important to note that in such a case, the airline will face a relatively high leasing payment that will be reflecting interest levels and depreciation that the carrier would have had to face in case of purchase. So the typical situation when airlines choose to rent their flight equipment is to find smaller depreciation and financial costs than when the equipment is purchased. In fact, North American airlines include the rental cost under the depreciation heading.

The cost of flight crew training when not amortizable is also included as a flight operation cost. Finally, other flight operation costs will include all those components related to the flight or the aircraft waiting time that could not be incorporated into other flight operation cost elements.

The second cost item considered as direct is *maintenance and overhaul*. This heading takes into account the costs of personnel directly or indirectly related to the maintenance activity, costs of subcontracted services, spare parts, components and so forth. In order to express it as an hourly basis it is divided by block hours flown per each type of plane.

The third and last component of direct costs is *depreciation* and *amortization*, being the normal practice the application of a linear amortization with a certain residual value. It may be translated as a cost per hour in a similar way as indicated above. It should be pointed out that ICAO considers ground property and equipment depreciation as a direct cost, which might be inappropriate unless such equipment were specific to a type of plane.

The first constituent of indirect costs, as ICAO advises, is *user charges and station expenses*. It includes airport charges like landing and parking fees, passengers and cargo charges, security charges, etc. It also incorporates en route charges that must be paid to the country whose air space the aircraft goes through. Landing and en route charges are commonly dependent on aircraft weight, hence, they should be given the consideration of direct rather than indirect costs. However, passengers and other airport charges do not have to do with type of plane, therefore, as they are not directly related to the aircraft they should be considered as indirect. ICAO finally includes all airport and en route fees under the indirect cost heading, though some of their components are clearly direct costs. Station expenses that arise as a result of the performance of handling services are given the consideration of indirect costs as well. It includes the handling personnel and ground equipment costs or payments for handling services contracted out.

Passenger service costs are also indirect. They include all those costs related to catering services and passengers comfort like cabin crew costs, meal costs, costs resulting from flight delays and cancellations, etc. *Ticketing*, *sales* and *promotion* items consider costs of the personnel involved in such activities, travel agencies commissions and advertising expenditures. Finally, the *general and administration* heading captures all those costs that cannot be allocated to a particular area, and when there are costs that could not be fitted inside any of the examined categories, they will be included in the *other cost* entry.

In addition to the distinction between direct and indirect cost, operating costs might be classified as fixed or variable depending upon their relationship with production. Air transport companies produce air services for passengers and cargo; however, when one tries to measure output, several alternatives for the definition of airlines' output exist. Available or performed passenger-kilometres, available or performed tonne-kilometres,[3] number of hours flown, number of kilometres flown or number of landings, are all alternative measures of airlines output. Even assuming that the selected measure of production is adequate, the question of what costs will remain fixed when output changes and what costs will vary with it, does not have a simple answer. Therefore, it might happen that some cost items could be immediately escapable if a flight is cancelled and hence would be clearly variable (i.e. fuel, passenger meals, etc.), while others could be partly escapable but not totally. For instance, cabin crew subsistence and overtime costs could be escapable if a flight is cancelled, though fixed salaries could not be. In addition, some costs that are clearly fixed in the short run will become variable when a longer perspective is adopted. The schedule program of services might vary from one term to another, and therefore, by changing flight plans some cost components might now be escapable. For example, aircraft might be sold and, as a result, the depreciation costs reduced.

Given the ICAO categorisation of costs, it is not possible to disentangle all cost items, and then it is very difficult to separate strictly what is escapable or not in the short or medium term. However, as Doganis (1995) points out, as much as 90% of total airlines costs can be varied in the medium term (a period of a year or so), either by discontinuing all operations or by a withdrawal of certain operations. This would be crucial for marginal cost calculations.

3.2. Review of Air Transport-Specific Literature and Reporting of Marginal Cost Estimates

Most studies dealing with marginal cost estimates in air transport utilise an econometric approach. The starting point is to choose an adequate functional form that will reflect the behaviour of costs in the industry, the translog specification being the most common. For instance, Caves, Christensen, and Tretheway (1984) try to explain the apparent paradox for US local carriers that seemed to have higher unit cost than trunk US carriers. The main novelty of this article is that it includes two dimensions of airline size: network size and passengers and freight transportation services provided; such distinction allows the authors to distinguish between returns to density (*RTD*) and returns to scale (*RTS*):

- *RTD* or variation of unit costs caused by increasing transportation services within a network of given size. More specifically, they define *RTD* as the proportional increase in output made possible by a proportional increase in all inputs, with point served, average stage length, average load factor, and input prices held fixed. This is equivalent to the inverse of the elasticity of total costs with respect to output (ε_y). *RTD* are said to be increasing, constant or decreasing, when *RTD* is greater, equal or less than unity. The authors use the terms increasing returns to density and economies of density interchangeably.

$$RTD = \frac{1}{\varepsilon_y}$$

- *RTS* or variation in unit costs with respect to proportional changes in both network size and the provision of transportation services. More specifically they define *RTS* as the proportional increase in output and point served made possible by a proportional increase in all inputs with average stage length, average load factor and input prices held fixed. This is equivalent to the inverse of the sum of the elasticities of total costs with respect to output (ε_y) and points served (ε_p). *RTS* are said to be increasing, constant or decreasing when *RTS* is greater, equal or less that unity. They use the terms increasing returns to scale and scale economies interchangeably.

$$RTS = \frac{1}{\varepsilon_y + \varepsilon_p}$$

The data set consists of annual observations for trunk and local US airlines during the period 1970–1981, with a total of 208 observations. These are aggregated data for the whole network. Five categories of inputs are employed: labour, fuel, flight equipment, ground property and equipment and other inputs. Finally, an aggregated output index is applied.

They find that there are substantial economies of density for all carriers, which means that unit costs decline with production for a specific airline markets. Such a finding confirms previous beliefs. They also find constant *RTS* for trunk airlines, but more importantly, they conclude that this finding applies also to local service airlines. Table 2 shows some costs elasticities reported by these authors. Table 3 presents *RTD* and *RTS* values.

Kirby (1986) follows a similar approach, however, he points out that Tonne-Miles Performed (*TMP*) is a traditional measure of output that ignores the multi-product nature of airlines' output, and that the same *TMP* aggregate output can be produced in different ways with important

Table 2. Cost elasticities[a] from Caves, Christensen, and Tretheway (1984).

Regressor	Unrestricted Translog Total Cost Function	Unrestricted Translog Variable Cost Function
Output	0.804	0.719
	(0.034)	(0.43)
Points served	0.132	0.139
	(0.031)	(0.033)
Stage length	−0.148	−0.046
	(0.054)	(0.055)
Load factor	−0.264	−0.145
	(0.070)	(0.071)
Labour price	0.356	0.422
	(0.002)	(0.002)
Fuel price	0.166	0.196
	(0.001)	(0.001)
Capital-materials price[b]	0.478	0.382
	(0.002)	(0.002)
Capacity		0.153
		(0.045)

[a]Extracted from Table 3 of original article. Standard errors in parentheses.
[b]Material price for the variable cost function.

Table 3. *RTD* and *RTS* values[a] from Caves, Christensen, and Tretheway (1984).

1970–1981	Total Costs		Variable Costs	
	RTD	*RTS*	*RTD*	*RTS*
Pooled mean	1.243	1.068	1.179	0.988
	(0.053)	(0.049)	(0.061)	(0.057)
Trunks mean	1.235	1.068	1.119	0.965
	(0.061)	(0.052)	(0.071)	(0.060)
Locals mean	1.254	1.069	1.265	1.019
	(0.073)	(0.077)	(0.083)	(0.085)

[a]Extracted from Table 4 of original article. Standard errors in parentheses.

implications for the airlines' costs. This difficulty has been met in the literature by the introduction of "output modifiers" like the average stage length (*ASL*), the average aircraft size (*AAS*) or the average load factor (*ALF*). The author considers these variables as ad hoc selected and not merely correlated with *TMP*, but directly linked through the following relationship:

$$TMP = PORTS \times ASL \times ALF \times AAS \times ADPP$$

where: *PORTS* is the number of airports served and *ADPP* the average number of departures per airport.

In this sense, correlation between scale and unit costs would be due to the collinearity between the level of output and the various network and technology variables, rather than to a direct causal relationship.

In order to avoid such a problem, the author works with an output index. Labour and fuel are the two input factors considered. Again, the translog specification for the cost function is chosen. The set of airlines selected consists of a group of Australian and US carriers observed during the period 1971–1978. A total of 148 observations are computed. Table 4 shows some of Kirby's results.

Kumbhakar (1990) builds on the previous paper of Caves, Christensen, and Tretheway (1984). The author extends their results in four ways:

- First, instead of a translog cost function he adopts a generalisation of the Symmetric Generalised McFadden (*SGM*) cost function that is flexible and globally concave in input prices.
- Second, he estimates the system of input demand functions derived from the cost minimising problem instead of the cost function. This allows to incorporate firm- and input-specific effects.
- Third, a general measure of exogenous technical progress is introduced.

Table 4. Cost elasticities[a] from Kirby (1986).

Output Dimension	Elasticity
PORTS	1.041
ASL[b]	0.911
ALF	0.314
AAS	0.495
ADPP	1.084
PASS	0.517
SCH	0.286
FPL	0.242
FPF	0.133

Note: *PASS*, passenger traffic; *SCH*, schedule services; *FPL*, price of labour; *FPF*, price of fuel.
[a]Extracted from Table 2 of original article and completed with other elasticities reported in the same article.
[b]Evaluated at 370 miles.

- Fourth, the data set is updated through 1984 and extended to include some new local service carriers. Hence the data set consists of a panel of US trunk and local airlines for the period 1970–1984.

His main findings are presented in Table 5.

Kumbhakar concludes that there is evidence of economies of density which has declined in the era of deregulation, which suggests that airlines have taken advantage of falling average cost by increasing output through more flights and denser seating arrangements. However, economies of density have not been fully exhausted. The estimates of *RTS* make the author not to accept the presence of constant *RTS*,[4] though *RTS* have declined after deregulation.

Gillen, Oum, and Tretheway (1990) continue also the work by Caves et al. (1984). The novelty of this article is the application of a multi-product approach to Canadian airlines. Again the main question is whether or not local carriers are subject to constant *RTS*.

The multi-product approach starts with the distinction among three types of airlines output: schedule passenger services, schedule freight services and charter services (passengers and freight).[5] The data set regards the Canadian airlines industry from 1964 to 1981, with a total of 117 observations. The preferred total cost model includes the three output variables, average stage length for schedule passenger services, number of points served by the carrier, three input prices (labour, fuel and other materials) and a time trend variable representing the technological changes. Obtained values of costs elasticities, *RTD* and *RTS* are reported in Tables 6 and 7.

The authors conclude that there are significant (unexploited) economies of density[6] at all but Air Canada's level of output, which indicates that smaller carriers have higher unit costs than the largest carrier, Air Canada.

Table 5. Cost elasticities *RTD* and *RTS*[a] from Kumbhakar (1990).

Elasticity	Full Model	Before Deregulation	After Deregulation	Trunk	Local
ε_y	0.7830	0.7317	0.8342	0.7368	0.7363
	(0.0277)	(0.0316)	(0.0315)	(0.0328)	(0.0354)
ε_p	0.0819	0.0798	0.0832	0.1043	0.1581
	(0.0253)	(0.0263)	(0.0271)	(0.0244)	(0.0448)
RTD	1.2771	1.3666	1.1987	1.3571	1.3581
	(0.0451)	(0.0590)	(0.0453)	(0.0604)	(0.0654)
RTS	1.1562	1.2321	1.0899	1.1889	1.1179
	(0.0426)	(0.0548)	(0.0435)	(0.0488)	(0.0676)

[a]Extracted from Tables I and III of original article. Standard errors in parentheses.

Table 6. Cost elasticities[a] from Gillen, Oum, and Tretheway (1990).

Variable	Total Cost Function	Variable Cost Function
Passenger	0.734	0.540
	(0.060)	(0.068)
Freight	0.048	0.039
	(0.041)	(0.053)
Charter	0.044	0.039
	(0.017)	(0.024)
Points served	0.204	0.308
	(0.099)	(0.137)
Labour price	0.322	0.374
	(0.008)	(0.009)
Fuel price	0.199	0.254
	(0.005)	(0.004)
Capital and material price	0.478	—
	(0.008)	
Materials price	—	0.371
		(0.009)
Stage length	−0.181	−0.112
	(0.056)	(0.129)

[a]Extracted from Table 2 of original article. Standard errors in parentheses.

Table 7. *RTD* and *RTS* values[a] from Gillen, Oum, and Tretheway (1990).

	Total Costs		Variable Costs	
	RTD	*RTS*	*RTD*	*RTS*
Sample mean	1.211	0.971	1.486	0.992
	(0.061)	(0.061)	(0.170)	(0.103)

[a]Extracted from Table 4 of original article and other values reported in the same article. Standard errors in parentheses.

On the other hand, the finding of constant *RTS*[7] suggest that a small network carrier should not have a cost disadvantage, provided it achieves traffic densities within its small network similar to those of Air Canada and Canadian Airlines International.

Windle (1991) measures productivity and unit cost for a group of US and non-US airlines. In this work he also reports the result of a previous work by Caves, Christensen, Tretheway, and Windle (1987), where a translog variable cost function is estimated. Labour, fuel and materials are the input

Table 8. Cost elasticities[a] from Caves, Christensen, Tretheway, and Windle (1987).

Variable	Variable Cost Function
Output	0.481
	(0.039)
Labour price	0.423
	(0.003)
Fuel price	0.200
	(0.002)
Stage length	0.006
	(0.044)
Load factors	−0.175
	(0.069)
Points	0.121
	(0.032)

[a]Extracted from Table 7 of original Windle (1991) article. Standard errors in parentheses.

factors applied; airlines' characteristics include: numbers of points served, load factors and stage length. The study period is 1970–1983. Elasticities values are shown in Table 8.

Brueckner and Spiller (1994) make use of airlines data at the level of the route, which may be regarded as an important step forward in the study of airlines' costs. Nevertheless, the use of disaggregated data to the route level requires, according to the authors, a methodological approach that differs from the traditional one (translog functions estimates). They rely on a structural model of competition among hub and spoke airlines, based on several assumptions regarding the nature of the oligopoly game played by the airlines as well as the functional forms of demand and marginal costs. The model yields two nonlinear reduced-form equations that relate fares and traffic levels in individual markets to traffic densities on the network spokes serving those markets. They show that marginal cost fall by 3.75% for every 10% increase in spoke traffic, pointing out that this density effects is stronger than that estimated by Caves et al. (1984), whose 0.80 cost elasticity with respect to traffic translates into a marginal cost elasticity of −0.20.

Oum and Zhang (1997) point out that traditional measure of *RTS* in transport overlook the role of operating characteristics, other than network size. They illustrate the link between operating characteristics such as load factor, stage length, network size and scale of outputs. By using the results obtained by Caves et al. (1984) and taking into account the link just mentioned, they find mildly increasing *RTS* for the airline industry.

Within the UNITE project, a case study on airlines costs was carried out as well (Betancor & Nombela, 2001). The basic data set consists of a panel of 22 airlines (13 European and 9 from North America) and nine consecutive years from 1990 to 1998. The aim of this case study was to estimate marginal costs for European airlines.

Two main methodologies were applied in order to perform marginal costs estimates. The first one is based on a disaggregated costs approach that distinguishes the four UNITE categories of costs (i.e. vehicle, service, infrastructure and administrative/commercial). In this case, the definition of marginal costs is the classical one:

$$MC = \frac{\Delta TC}{\Delta X}$$

where TC is total costs, which in turn could be the vehicle, service, infrastructure or administrative/commercial costs, and X is any airline output measure used.

The marginal costs was then estimated for each airline and for each regional sample, taking increments year by year since 1990 to 1998 and producing an average result for the whole period. Obviously, this simple approach assumes that changes in output measures are the only cause of changes in cost. It is important to test how realistic this assumption is. The second methodology selected applies econometrics techniques that allow to estimate a system of equations that include a translog cost function and a series of productive factors share equations. This method was more adequate in the sense that it took into account special features of each carrier as levels of activities, network size, stage length, factor prices and so on. However, it did not permit to separate by different type of UNITE cost categories, leading to a global marginal cost estimate related to all activities.

By combining both estimation methods we were in the position to test the robustness of marginal costs estimates. Therefore, the final value of marginal cost estimated in this work has the following features:

- It was the marginal cost of European airlines as a group.
- It was provided in terms of hours flown and tonne-km produced by airlines.
- Its robustness is based on the fact that two methodological approaches are applied.
- It was comparable only with equivalent concepts of revenue.

It is obvious that the marginal costs estimates should take into account only variable costs. However, this distinction has very much to do with the period of time considered. It has been already mentioned that as much as 90% of airline costs might be escapable in the medium term (Doganis, 1995). This is the reason why marginal costs are calculated by taking variation in all four UNITE categories of costs, though one might think that the costs included inside any particular heading will be more easily variable than those under other heading.

When implementing this first methodological approach, the way to proceed was the following:

(1) Calculate the yearly incremental variations of costs for each airline and UNITE cost category.
(2) Select measures of air carriers output and calculate yearly incremental variations. Output measures used were hours flown and available tonne-km.
(3) Calculate yearly marginal costs by dividing variation of cost by variations of output.
(4) Calculate an average with all valid marginal costs observations for each airline and cost component.

In this process two main problems appeared:

(1) Costs and output data for all airlines and for all years were not always available.
(2) Incremental variations were sometime negative. There appeared to be an efficiency effect that made the total costs function shift downwards or sometimes there were reductions in output over time.

To solve these troubles, observations that give rise to the problem were dropped from the sample in such a way that the final average of marginal costs would include only positive marginal costs that would result from positive incremental variation of both costs and outputs, with the requirement of having obtained at least two or more valid observations per airline.

Marginal and average cost estimates per each regional sample are shown in Tables 9 and 10. It can be seen that European airlines have always higher marginal and average costs than North American ones either in terms of hours flown or available tonne-km. Additionally, in both samples, marginal costs in terms of hours flown were higher than average costs, though very close to each other. On the contrary, the marginal cost estimate in terms of available tonne-km was always lower than the average cost, but again values were very similar. These results might be interpreted as evidence of the existence of constant *RTS* in the industry,[8] however, in order to reach a

Table 9. Marginal and Average Cost Estimates in Terms of Hours Flown (1998 €).

	Marginal Costs					Average Costs
	Vehicle Related (1)	Service Related (2)	Infrastructure Related[a] (3)	Adm/ Commercial Related (4)	All Categories	
European average (% over total)	3,935 (26.6)	2,182 (14.7)	3,681 (24.8)	5,017 (33.9)	14,815	11,278
North American average (% over total)	2,475 (34.9)	839 (11.8)	1,952 (27.5)	1,821 (25.7)	7,087	6,393

Note: (1) Average of British Airways, Finnair and SAS; (2) average of Air France, British Airways and Finnair; (3) average of British Airways, Finnair and SAS; (4) Finnair value.
[a]Includes handling costs.

Table 10. Marginal and Average Cost Estimates in Terms of Thousand Available Tonne-km (1998 €).

	Marginal Costs					Average Costs
	Vehicle Related (1)	Service Related (2)	Infrastructure Related[a] (3)	Adm/ Commercial Related (4)	All Categories	
European average (% over total)	265 (35.2)	98 (13.0)	162 (21.5)	228 (30.3)	753	838
North American average (% over total)	150 (39.2)	47 (12.3)	106 (27.7)	80 (20.9)	383	469

Note: (1) Average of British Airways, Finnair and SAS; (2) average of Air France, British Airways and Finnair; (3) average of British Airways, Finnair and SAS; (4) Finnair value.
[a]Includes handling costs.

definitive conclusion in this regard, we would need to make use of econometric means.[9]

When comparing weights for each cost component, it happens that these vary depending upon the type of output considered. In terms of hours flown, the most important cost category for European carriers is the administrative/commercial item, while for the North American airlines it will be the vehicle-related costs. When marginal costs in terms of tonne-km are considered, the vehicle-related appears now to be the most important over total marginal costs for both groups of airlines.

The marginal costs might also be calculated by estimating a function for airlines costs. This alternative allows to include some factors as average stage length, load factors or number of networks points, to control for airlines' characteristics. In addition, results provide relevant information about the industry, like costs elasticities, *RTS* and density, substitution elasticities between productive factors, and so on. The functional form selected was a translog, following other previous works.

Relevant variables and parameters are given by the following system of equations. Each variable description is shown in Table 11:

$$
\begin{aligned}
COST = {} & \alpha_0 + \alpha_1 OUTPUT + \beta_1 W + \beta_2 E + \beta_3 K + \beta_4 OM + 0.5\gamma_1 W^2 \\
& + 0.5\gamma_2 E^2 + 0.5\gamma_3 K^2 + 0.5\gamma_4 OM^2 + \gamma_5 WE + \gamma_6 WK + \gamma_7 WOM \\
& + \gamma_8 EK + \gamma_9 EOM + \gamma_{10} KOM + \delta_1 OUTPUTW + \delta_2 OUTPUTE \\
& + \delta_3 OUTPUTK + \delta_4 OUTPUTOM + \lambda_1 TLF + \lambda_2 ASL + \lambda_3 NET
\end{aligned}
$$

$$S_W = \beta_1 + \gamma_1 W + \gamma_5 E + \gamma_6 K + \gamma_7 OM + \delta_1 OUTPUT$$

$$S_E = \beta_2 + \gamma_2 E + \gamma_5 W + \gamma_8 K + \gamma_9 OM + \delta_2 OUTPUT$$

$$S_K = \beta_3 + \gamma_3 K + \gamma_6 W + \gamma_8 E + \gamma_{10} OM + \delta_3 OUTPUT$$

Except for the share of each factor price, all variables have been converted into logs to facilitate interpretation of coefficients as elasticities. Once logs were calculated, variables are transformed into deviations with respect to their mean values, for all but for ln(*COST*).

Total number of observations for which information on all variables existed reached a figure of 125 for the total sample. With the aim of getting an estimate of marginal cost for European airlines, the sample was split, resulting in 58 observations for Europe and the remaining 67 for North America. The Full Information Maximum Likelihood estimation method was the procedure applied to the whole sample, however, as the number of observations is very much reduced for the sub-samples, this method was no longer applicable, and SURE (Seemingly Unrelated Regression) was the alternative method selected. Since, by definition, the sum of factor shares is one, only three of the four share equations could be estimated simultaneously. Actually, better fits were obtained for the case in which the share equation of other materials (*OM*) was left aside. In addition, two alternative

Table 11. Description of Variables for Econometric Models.

COST	Total cost (million of 1998 €)
OUTPUT	Available tonne kilometres (thousand) or hours flown
W	Labour price or average salary per employee (thousand €)
E	Energy price or cost of energy per kilometre (€). It includes fuel and oil
K	Capital price or cost of capital per aircraft (million €). It includes depreciation, rental of flight equipment, maintenance and flight equipment insurance
OM	Other material price or cost per departure (thousand €)
WE, WK, WOM, EK, EOM, KOM	Cross products of factors prices
OUTPUTW, OUTPUTE, OUTPUTK, OUTPUTOM	Cross products between output and factors prices
TLF	Total load factor
ASL	Average stage length (kilometres per departure)
NET	Network points in 1997
S_W, S_E, S_K, S_{OM}	Factor shares of total costs

output variables have been used in order to obtain marginal cost estimates in terms of available tonne-km or number of hours flown.

Coefficients estimates are summarised in Table 12. From the lists of estimated coefficients, the most important for this work is α_1 that will be capturing the relationship between total cost and output. From this coefficient, a marginal cost estimate can be obtained for average values of all other variables by carrying out the following transformation:

$$MC = \frac{\partial COST}{\partial OUTPUT} = \alpha_1 \frac{COST}{OUTPUT} = \alpha_1 AC$$

where AC is the average costs that results from dividing mean values of *COST* and *OUTPUT* for each sample. The corresponding values of marginal costs are reported in Table 13.

By comparing Tables 9 and 10 with Table 13, it can be seen that both set of values are very close. This can be interpreted as a signal that previous estimates performed according to the methodology of a disaggregated costs approach would be robust.

Regarding other coefficients estimates and goodness of fit, it seems that econometric models estimated may be qualified as good. In general, all parameters and corresponding elasticities have the expected sign. It is important to note that these estimates not only allow to get, in turn, marginal cost

Table 12. Coefficient Estimates and Goodness of Fit.

	Total Sample (Obs = 125)		European Sample (Obs = 58)		North American Sample (Obs = 67)	
	Output: Available tonne-km	Output: Hours Flown	Output: Available tonne-km	Output: Hours Flown	Output: Available tonne-km	Output: Hours Flown
α_0	8.345*	8.348*	8.254*	8.505*	8.303*	8.333*
α_1	0.926*	0.935*	0.930*	1.015*	0.868*	0.880*
β_1	0.299*	0.300*	0.316*	0.287*	0.318*	0.311*
β_2	0.116*	0.116*	0.099*	0.095*	0.140*	0.140*
β_3	0.167*	0.167*	0.163*	0.164*	0.161*	0.164*
β_4	0.259*	0.583*	0.466*	0.335*	0.254*	0.435*
γ_1	0.227*	0.233*	0.222*	0.251*	0.261*	0.241*
γ_2	0.029*	0.048*	0.047*	0.052*	0.034*	0.050*
γ_3	0.096*	0.095*	0.086*	0.094*	0.093*	0.095*
γ_4	0.209	0.152	0.110	0.272*	0.396	0.142
γ_5	−0.043*	−0.016	−0.020*	−0.009	−0.011	−0.013
γ_6	−0.060*	−0.054*	−0.058*	−0.058*	−0.065*	−0.060*
γ_7	0.009	−0.017	−0.022	−0.002	0.003	−0.016
γ_8	−0.011*	−0.012*	−0.009	−0.005	−0.013*	−0.019*
γ_9	−0.019*	−0.024*	−0.018*	−0.014	0.006	0.001
γ_{10}	−0.067*	−0.064*	−0.060*	−0.064*	−0.068*	−0.060*
δ_1	−0.010	−0.019*	0.005	−0.023*	−0.033*	−0.029*
δ_2	0.014*	0.014*	0.007*	0.002	−0.003	−0.003
δ_3	0.005*	0.006*	0.001	0.002	0.012*	0.011*
δ_4	−0.100*	−0.030	−0.075*	−0.091*	0.069	0.043
λ_1	−0.451*	−0.164	−0.230	−0.659*	−0.360*	−0.357*
λ_2	−0.732*	−0.571*	−1.012*	−0.214*	−0.585*	−0.319*
λ_3	−0.128*	0.121*	−0.182*	0.089*	0.215*	0.157*
$COST$ R^2	0.981	0.987	0.986	0.992	0.981	0.992
S_W R^2	0.479	0.588	0.473	0.523	0.749	0.754
S_E R^2	0.633	0.470	0.723	0.609	0.384	0.290
S_K R^2	0.872	0.875	0.906	0.907	0.730	0.719

*Significant at 5% level of confidence.

values, they also may lead to factor prices elasticities, substitution elasticities, or to establish if return to scale and density are present or not.

Regarding economies of density (see Table 14), there appears to be constant or rather slightly increasing returns. Considering standard error values, the coefficient for the European sample could take values lower than 1, though very close to it. The presence of slightly increasing *RTD* appears to be more important in the case of North American carriers. These results

Table 13. Estimated Marginal Costs Resulting from Econometric Models.

	Total Sample		European Sample		North American Sample	
	€ per Thousand Available tonne-km	€ per Hour Flown	€ per Thousand Available tonne-km	€ per Hour Flown	€ per Thousand Available tonne-km	€ per Hour Flown
Marginal costs	473	6,921	644	12,255	399	5,596

Table 14. Returns to Density and Scale.

	Total Sample		European Sample		North American Sample	
	Available tonne-km	Hour Flown	Available tonne-km	Hour Flown	Available tonne-km	Hour Flown
Returns to density	1.079	1.069	1.075	0.985	1.152	1.136
Returns to scale	1.253	0.947	1.337	0.906	0.923	0.964

are consistent with marginal and average values obtained with the disaggregated cost approach.

RTS are quite different when the sub-samples are examined, leading to a result in which North American airlines would be experiencing diseconomies of scale. In fact, the coefficient λ_3, behaves differently for each sub-sample. However, this may be the consequence of the way in which the NET variable has been measured, as it was available only for the year 1997. Therefore, we cannot be conclusive in relation to the existence or not of *RTS*.

Regarding policy implication of results, several issues could be pointed out:

(1) Our results on marginal cost estimates have been calculated for European (or North American) airlines as a group, hence it is the marginal cost for an average European (North American) airline expressed in € per hour flown or per thousand available tonne-km taking into account their whole operations network. Such estimates might be compared with the average revenue (yield) of the group in order to reach some conclusions about allocative efficiency in the markets.

(2) The disaggregated cost approach allowed us to separate by type of cost, and to obtain marginal cost estimates for different cost groups (i.e. vehicle, service, infrastructure and administrative cost categories) in absolute and relative terms. This methodology also led us to estimates of marginal costs for some particular airlines that provided enough data. This was the case of Finnair for all four UNITE costs categories, and for British Airways, SAS and Air France in some cases. A better quality of data on costs provided by North American carriers resulted in more individual estimates of marginal costs for this group.

(3) The econometric models support values of marginal costs obtained by applying the disaggregated cost approach. However, its main drawback is its high level of aggregation. Again marginal costs estimates derived from this methodology are representative for each group of airlines, as it is the cost elasticity α_1. On the other hand, its main contribution regards the explanation of *RTD* in the industry. There appear to be constant or rather slightly increasing *RTD*. The presence of increasing *RTD* appears to be more important in the case of North American carriers. These results are consistent with marginal and average values obtained with the disaggregated cost approach.

(4) There is an important gap between North American and European Airlines costs, either in terms of marginal or average costs. Even taking into account the different economic environment in which the two group of carriers are operating, this result can be interpreted in the sense that there might be some scope for European airlines to reduce costs. However, in order to know the reasons behind such different cost behaviour, a more detailed analysis is needed. There is also another gap regarding provision of cost data. In the case of European airlines there is a general lack of regular yearly data on all ICAO categories. This does not occur with its North American counterparts.

4. RAIL TRANSPORT

4.1. Introduction

The first early on efforts to explain the nature and the behaviour of costs in transportation were made in the rail industry. This should not be a surprise taking into account that rail is one of the oldest modern modes. Lardner (1850) was one of the pioneer researchers on transport costs. Using the available statistics, this author produced a landmark work that shed light on the technology of laying down and maintaining the track. Steam locomotion was also studied from statistics by the same author, so that least-cost operation could be achieved. Before going any further on the detailed aspects of cost analysis in railways, a possible non-exhaustive categorisation of costs for this mode is presented in Table 15.

In cases of vertically integrated service providers, the identification of infrastructure-related costs and SOCs can present a very complex and not so clear cost structure. Other specific features of the railway industry have made the measurement of rail costs a fairly complex task. According to Braeutigam (1999) the following are some of the aspects leading to complexity:

- Railways service suppliers typically produced more than one type of freight service and many railways also provided significant amounts of passenger service.
- Some of the costs of a typical railway could be directly related with the production of individual services without any special problems (direct costs). However, other costs were shared in the production of two or more services. The shared costs (or common costs) were thought to represent a significant part of total costs (e.g. way and structure).
- Due to large fixed common costs, a railway would not be able to achieve financial viability if all services were priced at marginal cost. In such a case at least some rail rates would have to be higher than marginal cost to ensure financial break even.

Table 15. Operating Cost of Railway Operators.

Investment Costs			Costs for Operation and Maintenance			
Personnel Costs	Material Costs	Other Additional Costs	Personnel Costs	Energy Costs	Material Costs	Other Additional Costs
Rolling stock and facilities Installation Documentation			Costs for line operation			
Training of maintenance and operation personnel Initial spare package Maintenance infrastructure			Maintenance cost – preventive and corrective			

Source: Adapted from UNIFE (2001).

An additional element of complexity lies in the difficulty of identifying an element that can be used across the whole rail network as an allocation key to share common costs.

Typical concerns of early studies were how to allocate the different costs incurred in the production of railway services for purposes of pricing. Nowadays, much cost function research has to do with investigation of structural characteristics of the industry and efficiency comparisons across firms and/or over time (Oum and Waters, 1993). Nevertheless, these studies shed much light on the magnitude of marginal cost. The next section will present an overview of cost estimation studies in the rail sector with special emphasis in the marginal cost information produced in the UNITE research project.

4.2. Review of Rail Transport-Specific Literature and Reporting of Marginal Cost Estimates

As in other modes, there are three primary approaches for the study of costs and marginal cost estimation: econometric methods, engineering approaches and cost allocation based techniques.

In the econometric methods, costs are the dependent variable and transport outputs are among the independent variables. Cross-sectional and/or time-series analysis produce parameters that may be directly interpreted as marginal costs, or used to construct the total cost function. This type of analysis was applied in the rail sector before other modes, relating total operating cost per gross tonne-mile to gross tonne-mile per mile of track. From these initial formulations, it was recognised that traffic density was an important determinant of unit cost and that average length of haul affected cost-output relationships (Macário et al., 2003).

Until the 1970s the rail industry tended to adopt linear functions of the type

$$\frac{TOC}{MT} = a + b\left(\frac{GTM}{MT}\right) \tag{1}$$

where TOC represent the total operating costs, MT the miles of track and GTM the gross tonne-miles.

From the 1930s until 1980s, the former US Interstate Commerce Commission[10] had used these types of linear functions to shed light on the relation between cost and output. This regulatory body determined rates using fully allocated costs.

A meaningful break from linear cost functions was introduced by Keeler (1974). To derive and estimate log-linear cost functions for US passenger and freight operations, Cobb–Douglas production functions were used.

However, the major breakthrough in cost analysis studies was the introduction of less restrictive functional forms. These are commonly referred as flexible forms that include among others the generalised Leontief function, the quadratic mean of order R function, the generalised Cobb–Douglas function and the translog function.

The translog specification has proven to be the one most often employed in transport cost studies. One of the properties of translog functions is allowing for full modelling of substitution or complementarity between inputs.

Several studies have used econometric cost functions to test the existence of economies of scale and to compute marginal costs in the railway industry. The results are summarised in Table 16.

Table 16. Cost Studies Using Econometric Techniques.

Study	Functional Form	Data	Selected Focus or Conclusion
Keeler (1974)	Cobb–Douglas (loglinear)	Panel of 51 firms, 1968–1970	CRS, economies of density
Brown, Caves, and Christensen (1979)	Translog	Cross section, 67 firms, 1936	IRS
Caves et al. (1985)	Translog	Panel, 43 firms, 1951–1975	IRS for some carriers, CRS for large carriers, economies of density
Caves, Christensen, and Swanson (1987)	Translog	US railways, 1955–1974	Economies of density but less strong evidence of economies of scale, productivity of growth
Kim (1987)	Translog with Box-Cox	56 Class I railways in 1963	Slight IRS, diseconomies of scope between freight and pass
De Borger (1992)	Translog with Box-Cox	Belgium railways, 1950–1986	Constant returns
Friedlaender et al. (1993)	Translog	Panel, 1974–1986	Slight IRS, economies of density

Note: DRS, decreasing returns to scale; IRS, increasing returns to scale; CRS, constant returns to scale.

Source: Adapted from Braeutigam (1999).

It is clear that normally constant, or only slightly increasing *RTS* are found for the railway industry. However where *RTD* are also measured, strongly increasing *RTD* are found. *RTD* are measured by including both measures of traffic and of route kilometres into the equation, and calculating the effect on costs of an increase in traffic holding route kilometres constant. While *RTS* are relevant in considering the optimal structure of the railway industry in terms of the number and size of firms, *RTD* are the relevant concept for pricing purposes. Strongly increasing *RTD* indicate marginal cost well below average cost for rail services. However, since these studies all refer to vertically integrated firms it is not possible to tell whether this result arises from infrastructure or operations. The previous chapter has led us to expect increasing returns in infrastructure maintenance, but increasing returns may also apply in operations, for instance, from the operation of longer trains. The optimal adjustment of frequency and train length to demand is considered further in the following section.

A further interesting result is the presence of diseconomics of scope between passenger and freight operations in the study of Kim (1987). The results indicate that operating freight and passenger services together greatly increase costs compared with operating them separately. However, it is not clear whether this suggests structural change to the industry, with separate freight and passenger companies, or whether it is a consequence of shared use of the same track, with consequent conflicts between trains of different speeds.

Engineering approaches have also been applied to the study of railways costs. In these approaches, total costs are disaggregated into categories, and for each category, individual analysis provides the technical relationship between inputs and outputs and then relevant factors are applied to the resource requirements. Application of the engineering approach in rail passenger cost analysis is demonstrated in the report "User charges for railway infrastructure" (ECMT, 1998) and in the "Swedish rail case study"[11] included in the deliverable "SOCs case studies" of the UNITE project (Macário et al., 2003). In the latter the supplier marginal costs (carriage costs) of an interurban rail passenger transport service in Sweden are estimated. The analysis was carried out for a long-distance route between Stockholm and Sundsvall that has a total distance of 816 km,[12] operated with a flexible-formation train. Cost data from the manual for railway investment cost–benefit analysis issued by the Swedish Railtrack was used. The corresponding average marginal cost values are presented in Table 17. It will be seen that peak marginal costs are very much higher than off-peak, because in the peak adding carriages to trains requires purchase or

Table 17. Supplier Marginal Costs for the Interurban Rail Service Stockholm–Sundsvall.

	MC per Passenger km (Euro 1998)	MC per Passenger km in Peak[a] (Euro 1998)	MC per Passenger km in off Peak[b] (Euro 1998)
Swedish rail case study	0.036	0.072	0.022

Note: Considering an occupancy rate of 0.5, 80 seats per carriage and a voyage speed of 90 km/h.
Source: Macário et al. (2003).
[a]MC per passenger round trip in peak/line distance = 59/816 = 0.072 €.
[b]MC per passenger round trip in off-peak/line distance = 18/816 = 0.022 €.

leasing of additional rolling stock, whereas in the off-peak there is rolling stock spare.

Cost allocation approaches are based on general accounting systems and entail the allocation of accounting cost items to particular activities. This can be so simple as dividing total costs by total output level in order to get an average cost. Alternatively, more extensive regression analysis can be applied in order to get allocation factors in relation to distinct output measures (Bereskin, 1989). A practical example is the RFA (rail form A), used by the former ICC as a basis for the allocation of common costs and to separate fixed and variable costs. Much of the rail industry uses methodologies of this type.

5. URBAN TRANSPORT

5.1. Introduction

In the urban context, studies of operating costs have been carried out by several authors for mass transit rail (subway, tram and rail urban lines) and bus. A classification of urban transit systems is presented in Table 18.

In general, cost incurred in the production of the services can be divided into two broad categories:

- Capital – including facilities, vehicles, signalling systems and other equipment (automated ticket issuing and inspection systems).
- Operation – including labour (drivers, stewards, inspectors, ticket sellers, administration personnel, other staff), fuel/electric power, consumables (spare parts, lubricants, etc.), cleaning, routine maintenance and marketing.

Table 18. General Classification of Urban Transit Systems.

Paratransit[a]	Bus Transit	Rail Transit	Other
Taxi	Diesel bus	Commuter rail – operates on infrastructure that is generally part of a railways network for intercity passenger or freight trains.	Guided busways
Jitney	Trolley bus	Light-rail transit (also called streetcar or tram) – electric powered rail vehicles that share the road with other traffic.	Monorails
Ridesharing and Dial-a-bus, dial a ride.	Electric bus and Other power sources	Heavy-rail transport (also called subway, metro, underground, elevated or rail rapid transit)	Automated guideway transit – unmanned vehicles operating fixed guideways on exclusive rights of way.

[a]Demand responsive service.

In the past 20 years, urban transport operators, particularly bus operators, have showed a high interest in the analysis of costs at the micro level in order to adjust fares and service provision. According to Savage (1988, 1989) evolution in the cost analysis of urban bus systems has three main stages:

- Apportioning methods for allocating costs and revenues to route level were developed in the UK and USA in the period 1968–1974.
- Developments after 1974 mainly come from USA and Australia and were focused on prediction methods for incremental costs resulting from expanding/contracting service at particular periods of the day.
- After 1979, there has been a revival of interest in the UK with analysis of both allocated and incremental costs, particularly with the deregulation of the industry in 1986.

5.1.1. Review of Urban Specific Literature and Reporting of Marginal Cost Estimates

As for other modes, econometric methods, engineering based approaches and cost allocation-based techniques have been applied to interpret cost structures and cost behaviour of urban transport operations.

The basic concept behind cost allocation methods used in the study of transit service is that the cost of supplying the service is a function of the service produced, measured, for instance, in terms of vehicle-hours or kilometres or seat-kilometres.

An approach to supplier cost estimation, based on cost allocation, which relies extensively upon the analysis of cost information supplied by public transport operators was adopted by Allport (1981). This author developed a framework with a technical and time dimension applied to the study of bus, light and metro services. In the technical dimension, modal unit costs are related to five parameters: peak car requirement, vehicle hours, car kilometres, route kilometers and number of stops/stations. In the temporal dimension, the escapability of costs in the short, medium and longer term are defined, allowing quantification of marginal costs. The escapability concept is intrinsically related with the functional character of the activities attached to the different cost categories.

Within the UNITE project, a methodological approach based in an adapted version of the Allport's costing model was used to estimate the supplier operating marginal costs of a recent commuter heavy-rail service, crossing the Tagus river, in the Lisbon area (Macário et al., 2003).

The costing model used comprises two dimensions of analysis:

(1) A functional dimension, in which the SOCs are assigned according to a series of cost drivers: vehicle-km, train-km, passenger-km and peak vehicle per annum.
(2) A temporal dimension, in which the concept of escapability of costs is introduced, corresponding to the time allocation of the SOCs. The very short term is related to a possible scenario of instantaneous capacity increase, on a daily basis. It corresponds to short time periods where reinforcement of operational personnel shifts or small production adjustments may have occurred. In the short term, a capacity increase in terms of the number of size of vehicles must be provided. The long term is related to important capacity increases implying structural changes, including additional support staff, more logistic facilities or building of new stations.

Table 19. Supplier Marginal Costs for the Commuter Heavy-Rail Service Crossing the Tagus River in the Lisbon Area.

	MC per Train km in Peak	MC per Train km in Off-peak	MC per Passenger km in Peak	MC per Passenger km in Off-peak
Urban public transport case study, Lisbon	17.640	1.864	0.0216	0.0086

Note: Monetary values in € 1998.
Source: Macário et al. (2003).

The analysis of the accounting and output service data provided by the operator allowed for the calculation of marginal costs' proxy values (in this case average variable costs), using linear cost functions. The results are presented in Table 19.

Because average variable cost (*AVC*) was used as a proxy for marginal cost (*MC*), some theoretical aspects deserve to be highlighted:

- The average variable cost function measures the variable costs per unit of output. Starting in a zero level of output and considering the production of one unit of output, the *AVC* is just the variable cost of producing this one unit. If more units are produced one could expect that, at worst, *AVC* would remain constant. If production can be organised in a more efficient way, as the scale of output increases, *AVC* might even decrease initially. However, the fixed factors will eventually constrain the production process leading to the increase of *AVC* (for instance due to congestion).
- The marginal cost function measures the change in costs for a given change in output. One can present the definition of marginal cost in terms of the variable cost function:

$$MC(y) = \frac{c_V(y + \Delta y) - c_V(y)}{\Delta y}$$

This is the same to say that *MC* is the increase in total cost associated with a one unit increase in output. Then, when the firm is producing in a range of output where *AVC* are decreasing, it must be that the *MC* is less than the *AVC* in that range. Similarly, if the firm is producing in a range of output where AVC is rising, it must be that the MC is greater than the AVC. In short, $AVC = MC$ at the first unit of output and at the minimum of the AVC function. In the Lisbon case study, the necessary condition for $MC = AVC$ is that the operator is producing at a level of output where AVC is minimum. However, it is reasonable to assume that AVC is roughly constant, so that even if this condition does not hold, the results presented are a reasonable approximation to SRMC.

A similar approach, based on cost allocation, was applied in a case study developed within FISCUS research project (FISCUS D3, 1999) that allowed for the estimation of marginal vehicle costs for Lisbon bus transport. The results are presented in Table 20.

Early econometric studies used linear or exponential cost functions in the investigation of the behaviour of SOCs in urban bus transport systems. Evidence of constant RTS on bus systems has been reported by Koshal (1972), who used a linear cost function, and Nelson (1972), who applied an exponential cost function. Both studies were focused in the USA market and as such a different regulatory and organisational context affect the behaviour of cost drivers.

The study carried out by Koshal (1972) assumed that the total cost is a linear function of total bus mileage (representing total output). The cost function is as follows:

$$C = a + bM$$

where C is the total cost, M the total bus mileage, a the measure of economies of scale (if a is positive) or diseconomies of scale (if a is negative) and b the measure of the marginal cost ($b > 0$).

The value of the coefficient of cost elasticity, E_c, can be defined as follows:

$$E_c = \frac{C - a}{C}$$

Table 20. Marginal Vehicle-associated Costs of Lisbon Bus Transport.

	Short-Run Marginal Cost		Long-Run Marginal Cost	
(A) Passengers (B) Passenger-km	(A)	(B)	(A)	(B)
Lisbon buses – Carris	0.227	0.067	0.302	0.089
Lisbon underground	0.385	0.104	0.566	0.153

Note: Passengers and passengers-kilometre are in European Currency Unit (ECU) (1995).
Source: FISCUS D3 (1999).

The data used in the estimation, applying the least-squares method, refer to 1969 and relates to 10 bus corporations operating in US metropolitan areas. The marginal cost computed is US \$0.974. The fact that the constant term "*a*" is not significantly different from zero suggests the existence of constant *RTS*. Through the estimation of separate cost functions for operating costs and depreciation and maintenance the following partial marginal cost were obtained:

- Marginal cost of operation per bus mile = US \$0.829.
- Marginal cost of depreciation and maintenance per bus mile = US \$0.145.

A second generation of studies is characterised by the application of quadratic and translog cost functions.

Obeng (1984) developed and tested his model in 62 bus systems, randomly selected in USA, operating between 25 and 650 vehicles. The results of the short-term analysis show that smaller systems, operating less than 50 vehicles (representing 31% of the sample) presents diseconomies of density. Economies of density can be found in companies with higher dimension.

In a similar approach in 1997 the ISOTOPE study, a European funded research project, concluded that there are only limited economies of scale in the production of passenger bus services. The study showed that the optimum scale for bus operation would be 100 buses (ISOTOPE, 1997).

The form of the translog model used was the following one.

$$\begin{aligned}\ln C = {} & \alpha_O + \alpha_v \ln VK + \alpha_l \ln LK + \beta_l \ln P_l + \beta_k \ln P_k + \frac{1}{2}\delta_{vv}(\ln VK)^2 \\ & + \frac{1}{2}\delta_{ll}(\ln LK)^2 + \frac{1}{2}\gamma_{ll}(\ln P_l)^2 + \frac{1}{2}\gamma_{kk}(\ln P_k)^2 + \gamma_{lk} \ln P_l \ln P_k \\ & + \varphi_{vl} \ln VK \ln LK + \rho_{vl} \ln VK \ln P_l + \rho_{vk} \ln VK \ln P_k \\ & + \rho_{ll} \ln LK \ln P_l + \rho_{lk} \ln LK \ln P_k + \psi DV\end{aligned}$$

where C is the operating cost per annum, VK the vehicle kilometres per annum, LK the line kilometres per annum, P_l the price of labour, P_k the price of vehicles and DV the dummy variable (assuming value of 1 if city in Great Britain and 0 otherwise).

Another study by Savage (1997) estimates operating cost functions, also using translog specification, for urban mass transit rail systems in the USA. The data set covers the period 1985–1991 representing 13 heavy-rail (metro) and 9 light-rail urban (tramway) systems. The approach followed in this chapter involves the use of duality theory,[13] as described in Chapter 3.

As stated by the author, the transit system manager faces restrictions in the determination of all production factor quantities in the short run, namely, the degree of automation, quantity of infrastructure and number of vehicles. However, it is assumed that changes in the output might result in vehicles being transferred from the active to reserve fleets, with consequent impact in maintenance costs. Therefore, the short-run production function, which expresses the engineering relationship between physical input and outputs, was defined as follows:

$$Y = f(T^*, A^*, C, L, E)$$

where Y is the units of transit service, T the way and structure, A the automation, C the rolling stock, L the labour, E the propulsion electricity and the superscript * indicates that the quantity of output is fixed.

As it is assumed that the quantity of automation (A) and way and structure (T) cannot be changed in the short term, the short-run variable cost can be defined by

$$SRVC = P_{\text{cm}}C + P_{\text{l}}L + P_{\text{e}}E$$

where P_{cm} is the factor price of car maintenance, P_{l} the factor price of labour[14] and P_{e} the factor price of electricity.

Using the duality theory, the minimum cost function is derived from the production function. Therefore, minimising cost subject to a given level of output lead to the following expression:

$$SRVC = f(Y, H, T^*, A^*, P_{\text{cm}} + P_{\text{l}} + P_{\text{e}})$$

The general form of the cost equation estimated is represented as

$$\begin{aligned}\ln SRVC = {} & \Sigma_i \alpha_i \ln Y_i + \beta_i \ln T + \Sigma_i \gamma \ln H_i + \Sigma_i \theta_i D_i + \frac{1}{2}\Sigma_i \Sigma_j \alpha_{ij} (\ln Y_i)(\ln Y_j) \\ & + \frac{1}{2}\beta_{tt} (\ln T)^2 + \frac{1}{2}\Sigma_i \Sigma_j \gamma_{ij} (\ln H_i)(\ln H_j) + \Sigma_i \phi_{it} (\ln Y_i)(\ln T) \\ & + \Sigma_i \Sigma_j \phi_{ij} (\ln Y_i)(\ln H_j) + \Sigma_j \varphi_{tj} (\ln T)(\ln H_j) + \Sigma_j \lambda_i \ln P_i \\ & + \frac{1}{2}\Sigma_i \Sigma_j \lambda_{ij} (\ln P_i)(\ln P_j)\end{aligned}$$

where H_i is the continuous output characteristics and D_i the discrete output characteristics.

Two measures of output were used: car hours and load factor (calculated as passenger miles divided by revenue car miles). Economies of density (*ED*) are computed using the following equation:

$$ED = (\partial \ln SRVC / \partial \ln Y)^{-1}$$

The marginal cost of an additional car hour (*MC*), for each urban system, can be calculated using the product of the inverse of the point estimate of economies of density and the variable cost per car hour (*SRVC*). The expression is

$$MC = ED^{-1} \times SRVC$$

Values of *ED*, *SRVC* and *MC* are presented in Table 21, for each US transit system. It is clear that the marginal cost of an additional car is generally, though not always, somewhat below the average variable cost.

The marginal cost per passenger mile, generated by the marginal car hour, can also be computed if one divides the marginal cost of an additional car hour by the average passenger miles per car hour, for the system. This calculation (which obviously assumes constant load factors) was carried out by Savage (1997), who also compares these marginal cost estimations with the fare per passenger mile in order to draw general conclusion on pricing.

6. CONCLUSION

This chapter has reviewed the methodology and evidence on measurement of the marginal cost of public transport operations for all modes of transport, and conclusions may be drawn on both issues.

Regarding methodology, most of the studies we reviewed used either the cost allocation or econometric approaches, with more limited use of the engineering approach. The strength of the econometric approach is that it provides firm evidence on the relationships, with statistical tests of significance and confidence intervals, with much less reliance on prior assumptions than the cost accounting approach, particularly where a flexible functional form such as the translog is used. It provides evidence therefore, on which variables influence costs and on the shape of the relationship, and in particular the extent of the existence of economies of scale, density and scope.

The disadvantage of the econometric approach is that it generally only provides aggregate measures; it cannot usually provide detailed estimates of

Table 21. Economies of Density and Marginal Costs in US Rail Transit System.

Rail Urban Mass Transit Systems	Data[a]	Type of System	Economies of Density[b] – Car Hours and Load Factor	Variable Cost per Car Hour (US Dollars – Constant Prices 1991)	Marginal Cost of an Additional Car Hour[c] (US Dollars – Constant Prices 1991)
Newark	1985/91	Streetcar	3.13	80	25.56
Cleveland	1986/91	Traditional light-rail	1.72	175	101.74
Portland, Oregon	1988/91	New light-rail	1.34	120	89.55
Buffalo	1988/91	New light-rail	1.09	114	104.59
Cleveland	1986/91	Traditional heavy-rail	1.60	175	109.38
New Orleans	1987/91	Streetcar	1.04	51	49.04
New York (Staten Island)	1988/91	Traditional heavy-rail	2.30	141	61.30
Baltimore	1990/91	New heavy-rail	4.70	140	29.79
Pittsburgh	1988/91	Traditional light-rail	1.21	110	90.91
Philadelphia (Lindenwold)	1985/91	New commuter heavy-rail	7.81	123	15.75
Miami	1990/91	New commuter heavy-rail	1.05	181	172.38
San Diego	1985/91	New light-rail	0.97	62	63.92
San Francisco (MUNI)	1986/91	Traditional light-rail	1.05	137	130.48
Philadelphia (SEPTA)	1985/91	Traditional light-rail	1.06	81	76.42
Atlanta	1985/91	New heavy-rail	1.23	75	60.98
New York (PATH)	1986/91		2.59	188	72.59

Table 21. (*Continued*)

Rail Urban Mass Transit Systems	Data[a]	Type of System	Economies of Density[b] – Car Hours and Load Factor	Variable Cost per Car Hour (US Dollars – Constant Prices 1991)	Marginal Cost of an Additional Car Hour[c] (US Dollars – Constant Prices 1991)
		Traditional heavy-rail			
Philadelphia (SEPTA)	1986/91	Traditional heavy-rail	0.63	113	179.37
Boston	1986/91	Traditional heavy-rail	1.08	144	133.33
San Francisco (BART)	1985/91	New commuter heavy-rail	1.91	117	61.26
Washington, DC	1986/91	New heavy-rail	1.39	133	95.68
Chicago	1985/91	Traditional heavy-rail	3.86	96	24.87
New York Transit Authority	1985/91	Traditional heavy-rail	0.98	113	115.31

Source: Savage (1997).

[a]Cost data used refers to "total mode expense" less "non-vehicle maintenance". Capital or expenditure charges are not included.

[b]Point estimates for individual systems.

[c]Own calculation using data and results presented in Savage (1997). Does not include the purchase of additional rolling stock or the effects of additional car hours on way and structure wear and tear.

the costs for specific services. This is both because the data available is typically aggregate (for a firm as a whole rather than an individual route) and the number of different output measures that can be introduced in the equation is limited by the high degree of collinearity between them. Thus, for instance, in the econometric studies of air transport, the only variables used to distinguish types of service are very aggregate ones such as trunk versus local, stage length or load factor. Type of vehicle can only be distinguished in terms of gross tonne kilometres, and no econometric study of which we are aware produced different results for peak and off-peak marginal costs. This means that the cost allocation approach, informed by the econometric results, has an important place as well. The evidence from the econometric studies is that public transport operations generally are produced under conditions approximating to constant *RTS*, although there is some evidence of increasing returns with respect to traffic density. The assumption of constant *RTS* that typically forms part of the cost allocation approach may therefore be reasonable. By distinguishing between costs related to numbers of vehicles required, hours and kilometres operated, the cost allocation approach permits cost to be distinguished by time of day, and by route according to its mean operating speed. The evidence presented here suggests that for interurban rail services peak costs are approaching four times off-peak, and in the urban rail case study the ratio is nearer to eight. Thus, for practical pricing decisions the distinction between peak and off-peak marginal costs is very important.

NOTES

1. The extent of evidence relating to maritime transport is somewhat less than for rail, air and urban transport and it was decided, therefore, to omit coverage of maritime transport. However, the interested reader is referred to Jansson and Shneerson (1982), Stopford (1997), Talley, Agarwal, and Breakfield (1986).
2. After the terrorist attacks on September 11th in Washington and New York, this item has increased substantially.
3. Tonne-kilometres is an aggregated passenger-cargo measure of total airline output. Each passenger is assigned 100 kg, of which 80 correspond to the person weight and 20 to luggage.
4. This finding contradicts the results of Caves et al. (1984).
5. Other possible levels of disaggregation of airlines' output are reported in Antoniou (1991).
6. These authors now use the term returns to traffic density for returns to density.
7. These authors now use the term returns to firm/network size for returns to scale.

8. Strictly speaking, one could find slightly increasing or decreasing returns to scale, depending on the definition of output used.

9. See below.

10. Former independent agency of the US government, established in 1887, that was in charge of the regulation of rail sector, which was terminated at the end of 1995.

11. In this case study three classical works, describing the engineering approach to production and cost functions, are referred: (1) Chenery, H. (1953). Process and production function from engineering data. In: W. Leontief (Ed.), *Studies in the structure of the American economy*. Oxford; (2) Smith, V. (1959). The theory of investment and production. *Quarterly Journal of Economics, 73*; (3) Walters, A. (1963). Production and cost functions: An econometric survey. *Econometrica, 31*.

12. This route has three segments: Stockolm/Gävle; Gävle/Söderhamn and Söderhamn/Sundsvall.

13. This theory assumes cost minimisation, which cannot be entirely realistic to many transport systems, where objectives may be different from the traditional profit maximisation (Nash, 1978). However, reductions in state aid have made it increasingly important for transport operators to adopt cost-minimising policies in service provision. Thus, assuming that operators choose to focus on labour, electrical power, fuel, vehicles and other materials to minimise the production cost at a given level of output, duality theory can be applied to transit systems. Demonstration on the use of dual theory can be consulted on McFadden (1978).

14. The author assumes that the price of labour is exogenous, but also refers that this might not be completely realistic due to the common belief that management have acquiesced with the unions to fix wages above those for comparable jobs. Note that duality theory assumes that factor prices are exogenous.

REFERENCES

Allport, R. J. (1981). The costing of bus, light rail transit and metro public transport systems. *Traffic Engineering and Control, 22*(December), 633–639.

Antoniou, A. (1991). Economies of scale in the airline industry: The evidence revisited. *Logistic and Transportation Review, 27*(2), 159–184.

Bereskin, C. G. (1989). An econometric alternative to URCS (Uniform Railroad Costing System). *Logistics and Transportation Review, 25*(2), 99–128.

Betancor, O., & Nombela, G. (2001). European Air Transport Operating Costs. UNITE Project Workpackage 6, Supplier Operating Costs, Case Study 6c.

Braeutigam, R. (1999). *Learning about transport costs. Essays in transportation economics and policy: A handbook in honour of John Meyer J. Gomez-Ibanez, W. Tie and C. Winston.* Washington, DC: The Brookings Institution.

Brown, R. S., Caves, D. W., & Christensen, L. R. (1979). Modelling the structure of cost and production for multi-product firms. *Southern Economic Journal, 46*, 256–273.

Brueckner, J. K., & Spiller, P. T. (1994). Economies of Traffic Density in the deregulated Airline industry. *Journal of Law and Economics, XXXVII*, 379–415.

Caves, D. W., Christensen, L. R., & Swanson, J. A. (1987). Productivity growth, scale economics and capacity utilization in US railroads, 1995–1974. *American Economic Review*, *71*, 994–1002.

Caves, D. W., Christensen, L. R., & Tretheway, M. W. (1984). Economies of density versus economies of scale: Why trunk and local service airlines costs differ. *Rand Journal of Economics*, (Winter), 471–489.

Caves, D. W., Christensen, L. R., Tretheway, M. W., & Windle, R. J. (1985). Network effects and the measurement of returns to scale and density for U.S. railroads. In: A. F. Daughety (Ed.), *Analytical Studies in Transport Economics*, (pp. 97–120). Cambridge: Cambridge University Press.

Caves, D. W., Christensen, L. R., Tretheway, M. W., & Windle, R. J. (1987). An assessment of the efficiency effects of us airline deregulation via an international comparison. In: E. E. Bailey (Ed.), *Public regulation: New perspectives on institution and policies*. Cambridge, MA: MIT Press.

Chenery, H. (1953). Process and production function from engineering data. In W. Leontief (Ed.), *Studies in the structure of the American economy*. London: Oxford University Press.

De Borger, B. (1992). Estimating a multiple-output generalized Box-Cox cost function : Cost structure and productivity growth in Belgian railroad operations, 1950–1986. *European Economic Review, Elsevier*, *36*(7), 1379–1398.

Doganis, R. (1995). *Flying off course. The economics of international airlines*. London: Harper Collins.

ECMT (1998). User charges for railway infrastructure. Round Table 107, Paris.

FISCUS (1999). D2 – Report on methodological framework to evaluate real transport costs, European Commission.

FISCUS (1999). D3 – Guide for the evaluation of real transport costs, European Commission.

Friedlaender, A. F., Berndt, E. R., Chiang, J. S., Showalter, M., & Vellturo, C. A. (1993). *Journal of Transport Economics and Policy*, *27*(2), 131–152.

Gillen, D. W., Oum, T. H., & Tretheway, M. W. (1990). Airline cost structure and policy implications. A multi-product approach for Canadian Airlines. *Journal of Transport Economics and Policy*, *24*(January), 9–34.

ISOTOPE. (1997). *Improved structure and organisation for urban transport operations of passengers in Europe*. Luxembourg: European Commission.

Jansson, J. O., & Shneerson, D. (1982). The optimal size ship. *Journal of Transport Economics and Policy*, *16*(3), 217–238.

Keeler, T. E. (1974). Railroad costs, returns to scale and express capacity. *Review of Economics and Statistics*, *56*(May), 201–208.

Kim, H. Y. (1987). Economies of scale and scope in multiproduct firms: evidence from US railroads. *Applied Economics*, *19*, 733–741.

Kirby, M. G. (1986). Airline economics of scale and Australian domestic air transport policy. *Journal of Transport Economics and Policy*, *20*(September), 339–352.

Koshal, R. K. (1972). Economies of scale in bus transport: Some United States experience. *Journal of Transport Economics and Policy*, *6*(2), 151–153.

Kumbhakar, S. C. (1990). A re-examination of returns to scale, density and technical progress in the US Airlines. *Southern Economic Journal*, *57*(2), 428–442.

Lardner, D. (1850). *Railway economy: A treatise on the new art of transport, its management prospects and relations*. New York: Augustus M. Kelley.

Macário, R., Carmona, M., Crespo Diu, F., Betancor, O., Nombela, G., Ericsson, R., & Jansson, J. O. (2003). Supplier operating costs case studies. UNITE (UNIfication of accounts and marginal costs for Transport Efficiency) Leeds, ITS. Funded by the 5th Framework RTD Programme.

McFadden, D. L. (1978). Cost revenue and profit functions. In: M.A. Fuss, & D. McFadden (Eds.), *Production economics: A dual approach to theory and applications*. Amsterdam: North Holland.

Nash, C. (1978). Management objectives, fares and service levels in bus transport. *Journal of Transport Economics and Policy*, *12*, 70–85.

Nelson, G. R. (1972). An econometric model of urban bus transit operations. Paper P-863, Institute for Defense Analysis.

Obeng, K. (1984). The economics of bus transit operations. *Logistics and Transportation Review*, *20*(1), 45–65.

Oum, T. H., & Zhang, Y. (1997). A note on scale economies in transport. *Journal of Transport Economics and Policy*, *31*(3), 309–315.

Oum, T. H., & Waters, W. G. (1993). A survey of recent developments in transportation cost function research. *Logistics and Transportation Review*, *32*(4), 423–460.

Savage, I. (1988). The analysis of bus costs and revenues by time period. *Transport Reviews*, *8*(4), 283–299.

Savage, I. (1989). The analysis of bus costs and revenues by time period. *Transport Reviews*, *9*(1), 1–17.

Savage, I. (1997). Scale economies in United States rail transit systems. *Transportation Research – A*, *31*(6), 459–473.

Smith, V. (1959). The theory of investment and production. *Quarterly Journal of Economics*, *73*, 61–87.

Stopford, M. (1997). *Maritime economics*. London: Routledge.

Talley, K., Agarwal, V. B., & Breakfield, J. W. (1986). Economies of density of ocean tanker ships. *Journal of Transport Economics and Policy*, *20*(1), 91–99.

UNIFE LCC Working Group. (2001). Guidelines for Life Cycle Cost – Volume II – Terms and definitions for total railway systems. Retrieved September 16, 2003, *from UNIFE – Union of the European Railway Industries – web site*: http://www.unife.org/workgroups/lcc_2.htm

Walters, A. (1963). Production and cost functions: An econometric survey. *Econometrica*, *31*(Jan–April), 1–66.

Windle, R. J. (1991). The world Airlines. A cost and productivity comparison. *Journal of Transport Economics and Policy*, *25*(January), 31–49.

USER COSTS AND BENEFITS

Claus Doll and Jan Owen Jansson[1]

1. INTRODUCTION

User costs and benefits describe the effects users mutually impose on each other when carrying out activities jointly. These interferences may either be of positive or of negative nature. Positive impacts occur in such situations, where additional public transport users require more capacity in the form of more vehicles, which would reduce the cost of access for existing users. In transportation this effect was first described and quantified by Herbert Mohring for urban bus transit (Mohring, 1972) and thus it is commonly entitled as the *Mohring effect*.

The negative impacts of additional vehicles in a given transport infrastructure system are more frequent in transportation than the positive ones. According to the final report of the expert advisors (Link & Maibach, 1999) to the High Level Group on Transport Infrastructure Charging, set in by the European Commission (Blonk, 1999), these can be classified into *congestion* and *scarcity effects*. Congestion occurs at infrastructure facilities where access is not regulated and thus demand can increase up to and beyond the infrastructure capacity. In this case the performance of the system, which might be measured in travel speed, goes down and we end up in congestion. Scarcity effects, on the other hand, occur in regulated systems where users cannot get a desired slot due to the presence of other users. As can be seen in air transport, both effects are not exclusive. Due to the non-availability of slots, operators may be forced to use second-best slots only. And even then, air traffic congestion might cause additional delays.

Measuring the Marginal Social Cost of Transport
Research in Transportation Economics, Volume 14, 125–154
Copyright © 2005 by Elsevier Ltd.
All rights of reproduction in any form reserved
ISSN: 0739-8859/doi:10.1016/S0739-8859(05)14005-0

Scarcity costs require estimates of the social value of alternative use of the slots, and are not considered further in this chapter. They remain a priority for future research.

For pricing policies, which is the main focus of the UNIfication of accounts and marginal costs for transport efficiency (UNITE) research, the external user effects, i.e. those costs or savings caused by an additional user to all other users, is relevant. From a mathematical perspective there is no fundamental difference between computing the positive or the negative user externalities. The basic formula for quantifying marginal external social congestion costs and Mohring benefits at a given traffic level Q is

$$MC_{\text{Ext}}(Q) = \frac{\partial TC(Q)}{\partial Q} - AC(Q) = Q\frac{\partial AC(Q)}{\partial Q} \quad (1)$$

where $TC(Q)$ is the sum of operator and user costs, $MC_{\text{Ext}}(Q)$ denotes the marginal external costs and $AC(Q)$ the average costs per user at the traffic level Q. It is obvious that in case of a positive relationship between the total costs and the traffic volume the marginal external costs are positive, while in case of falling total costs the expression gets negative and thus we end up in a social benefit. The two sides of the user costs and benefits effect and their consideration within the UNITE project will be presented in the subsequent sections.

2. THE COST SIDE: ESTIMATING CONGESTION COSTS

Valuing congestion cost involves three driving factors: the valuation of travel time, the relationship between travel time and traffic demand and the transport demand function. The value of time (VOT) and the time–demand relationship ($TT(Q)$) determine average, and thus marginal external user costs at the current traffic level Q is as follows:

$$MC_{\text{Ext}}(Q) = -Q\frac{VOT}{TT(Q)^2}\frac{\partial TT(Q)}{\partial Q} \quad (2)$$

As demand will react on the internalisation of the external congestion costs and thus the external congestion costs themselves will decrease, the optimal congestion charge will be in the equilibrium of the user cost curve, which is given by the sum of average costs and marginal external costs, and the demand curve. Thus, the demand function determines the relationship between actual external congestion costs and equilibrium congestion charges.

Each of these driving factors is of course specific to the preferences and to the behavioural pattern of each individual user and thus the exact description of the congestion phenomenon would require the application of very complex traffic simulation models. As such models are rather complex and costly in application, aggregated values or functions are commonly used to describe congestion.

2.1. The Valuation of Travel Time

The money value travellers or shippers put on time spent in traffic, or better on increases or savings of travel time, can be quantified by several very different procedures. First of all it is possible to think about the worth of time itself for the society. In this case, we would distinguish between work time and leisure time spent for transportation purposes. For work time, it is argued that the wage rate reflects the output that could be achieved in the time in question. For leisure time, studies of willingness to pay from (revealed preference) actual or hypothetical (stated preference) choices are the usual approach. With this approach, the German Empfehlungen für Wirtschaftlichkeitsuntersuchungen an Straßen manual (FGSV, 1997) arrives at a value for work hours of 9.92 €/person-hour and a value of non-working time of 1.44 €/person-hour. After considering the share of trips for work and non-work purposes and the respective vehicle occupancy rates the calculation arrives at values of 5.52 €/car-hour for weekdays and 281 €/car-hour at weekends and holidays.

However, it may be argued that the willingness to pay approach is needed for working time as well. Good examples for the superiority of values of resulting from willingness to pay (WTP) surveys are the valuations of freight and air travel time. In the case of freight transport it is argued, that besides the relatively easy computable costs of drivers, vehicles, fuel and other resource costs, the valuation of delays by the shipper's customers play an important role. The quantification of these preferences is hardly possible on the basis of entrepreneurial accounts figures, and thus the preferences of the shipping industries should better be taken into account by revealed or stated preference surveys. A comparison of the German values of time for goods vehicles, which are based on the macroscopic approach (30 €/hour for articulated trucks), and the ones based on U.K. WTP surveys and which are recommended by the expert advisors of the high-level group (42 €/hour for heavy trucks) suggests, that the valuation of the shippers' customers accounts to roughly 1/3 of total time costs. In the case of air traffic, or business travel in general, the departure and/or the arrival time and thus the possibilities of making use of the remaining time of the day of travel or the possibilities of avoiding overnight stays directly affects the travellers' valuation of time savings. These influences also cannot be captured by the top-down approach.

Users' preferences may be assessed by revealed or by stated preference surveys. Revealed preferences face the problem, that the survey design most likely does not capture each influencing factor and thus the selection of sample data will have a great impact on the study results. On the other hand, stated preference surveys might be influenced by mistrust of the interviewed people concerning the objectives of the survey and their inability to appraise their preferences. However, Nash and Sansom (1999) state that there is evidence that advanced interview techniques can well cope with these problems and thus that commonly stated preference surveys are superior to revealed preference studies.

In the UNITE project, three recent studies on the valuation of travel time were selected in order to receive a consistent valuation concept for the user cost and benefit case studies. These cores studies are:

- The U.K. value of time (VOT) study carried out by the Hague Consult Group in 1994 (Gunn & Rohr, 1996).
- The 1995/1996 Value of Time study in the Netherlands, conducted by the Hague Consult Group (Gunn, Tuinenga, Cheung, & Kleijn, 1999).
- The Swedish VOT study commissioned by the Swedish Institute for Transport and Communication Analysis (SIKA) and financed by the Swedish Communication and Research Board, the National Road Administration and the National Rail Administration (Algers, Hugosson, & Lindquist-Dillén, 1995).

The British VOT study has compiled a large number of surveys on the time–cost relationship using stated preference, revealed preference and transfer price techniques. The modes considered were car and lorry. For cars the values are given in pence per person-hour, where drivers and passenger and further commuting, other non-work trips, employers' and employees' travel are distinguished. In road haulage several lorry weight classes and the ownership of the vehicle are considered. Public transport was not reported. The study has shown, that the data used for stated preference (SP) studies did derive VOTs, which were consistent with revealed preference (RP) results, which adds to the evidence, that all three techniques are applicable to estimate the value of time, at least in the transport sector.

The Dutch VOT study 1997, which was commissioned by the Rijkswaterstaat uses stated preference and revealed preference data collected between 1990 and additional 5,000 interviews, which are used to update the 1990–1996 data to 1997. The transport segments considered by the study are private trips by car and by train and business trips, including all modes. Freight transport by either mode or air transport was not considered.

The methodology applied by the Swedish VOT-study was decided based on the conclusions of a Scandinavian Value of Time Conference, held in 1991 in Helsinki. It mainly follows the methods applied by the U.K. and the Dutch studies, and thus the values for private trips are estimated by the neoclassical model of individual utility maximisation under budget restraints. For business trips the productivity aspect of the individual is discussed but ignored due to its complexity. The modes considered are car, air, IC-train, regional train, coach and regional bus. Freight transport was not included.

Further information has been included on the values of freight transport of various modes (de Jony, 1996) and on air travel found by the project EUNET (Socio-Economic and Spatial Impacts of Transport Infrastructure Investments and Transport System Improvements) (Nellthorp, Mackie, & Bristow, 1998). The results of the different studies were compiled by transport sector and were updated to 1998. Table 1 presents the comparison and the selected values for the UNITE study.

For passenger transport by road and rail the values selected for use in the UNITE project seem to be rather reliable as they are confirmed by many empirical studies. However, the values recommended for air travel time lack empirical evidence as the two sources available report values of a different order of magnitude. In this field, further research should be carried out in order to get a better understanding of the driving factors of travel time values in aviation.

While in the freight sector the values for road haulage also are well investigated, rail and waterborne shipments are characterised by very different loading factors of trains or ships. Thus, the value of time could only be expressed by Euro per ton and hour. In all modes the type of the goods was not considered. This is not problematic as UNITE did not carry out in-depth case studies in scheduled freight transport. However, it is recommended that further research is dedicated to the specificities of these market segments.

2.2. Speed-Flow Functions

In road transport, speed-flow functions describe the physical relation between the number of vehicles competing for road space in a particular time

Table 1. Comparison of Selected VOT Studies (in Euro 1998).

Transport Segment	HCG 1994 U.K. 1994	HCG 1998 NL 1997	SIKA SE 1996	EUNET EU 1995	UNITE EU 1998
Passenger transport – VOT per person-hour					
Car/motorcycle		6.70	9.31		
Business	21.23	21.00	11.95		21.00
Commuting/private	5.53	6.37	3.91		6.00
leisure/holiday	3.79	5.08	3.10		4.00
Coach (Inter-urban)			7.47		
Business	21.23				21.00
Commuting/private	5.95		5.40		6.00
leisure/holiday	3.08		4.37		4.00
Urban bus/tramway			5.75		
Business	21.23				21.00
Commuting/private	5.95		4.94		6.00
leisure/holiday	3.08		3.22		3.20
Inter-urban rail		4.97	8.50		
Business		18.43	11.95		16.00
Commuting/private		6.48	6.21		6.40
leisure/holiday		4.41	4.94		4.70
Air traffic				40.60	
Business			16.20		16.20
Commuting/private			10.11		10.00
leisure/holiday			10.11		10.00
Freight transport – VOT per vehicle, train, wagon, ship and ton-hour					
Road Transport	36.00				32.60
LGV	45.00		39.68	30.75	40.76
HGV	48.00		39.68	30.75	43.47
Rail transport					
Full train load	801.00			645.37	725.45
Wagon load	32.00			26.16	28.98
Average per ton	0.83				0.76
Inland navigation					
Full ship load	222.00			178.55	201.06
Average per ton	0.20				0.18
Maritime shipping					
Full ship load	222.00			178.55	201.06
Average per ton	0.20				0.18

segment and the resulting travel speed. For a given traffic level, average user costs then are computed as the quotient out of the value of time and the travel speed per user group. We have identified four types of speed-flow curves:

(1) *Composed hyperbolic function* with "soft" capacity limit. The basic form of these functions is similar to the LOGIT-functions, however, the transition from free flow traffic conditions to congestion is defined more accurately by composing the speed-flow relationship out of several partially defined functions. In the German recommendations for assessments of road investment projects (FGSV, 1997) three partial functions are used.
(2) *Simple hyperbolic functions* take account of the fact that the impact of increasing traffic density is very minor for low traffic volumes and gets more and more obvious when traffic density increases. To describe this characteristic, hyperbolic or polynomic expressions for $v(Q)$ may be used. The official German speed-flow functions published in the Guidelines for road construction (FGSV, 1986) and applied in the General Investment Plan 1992 are of this nature.
(3) *LOGIT functions*, which describe travel speed in a single S-shaped curve. This function consists of a particular traffic volume, where traffic speeds decrease rapidly, while remain relatively stable before and after this point. LOGIT-type functions are, e.g. applied in the TRENEN-II-STRAN (Models for Transport Environment and Energy-Version 2-Strategic Transport Policy Analysis) project (Proost & Van Dender, 1999) in order to describe urban travel flows.
(4) *Linear relationships*, which describe the development of travel speed either by a single or by a set of partially defined linear functions of traffic volume of the form $v = aQ$. Partially defined linear functions are used in U.K. traffic planning (COBA-Manual on Cost-Benefit Analyses, Department for Transport, 2002). Simple linear curves have been proposed for the evaluation of congestion costs within the project PETS (Pricing European Transport Systems) (PETS, 1997).

Besides these basic functional definitions, speed-flow curves might be designed to end at zero travel speed or at a defined minimum travel speed. The decision, which is alternatively appropriate depends on the time interval considered. While in case of very short observation times travel speeds may well be zero, e.g. in the case of real congestion, for longer time intervals we must constitute that vehicles will move on and thus a minimum speed is required. However, as the minimum speed property says, that beyond a certain traffic volume average user costs are not affected by the arrival of additional users, marginal external congestion costs in this case are zero, while the traffic load on the link does not face a definite capacity limit. This is clearly not realistic.

The simplification of assuming a constant positive minimum speed is particularly relevant for single link models and is useful to ensure a stable run of network models for arbitrary demand scenarios. But it does not explain user reactions on a microscopic level and does not provide correct values of short-run marginal social costs. Fully specified demand models will account for the share of demand, which is choked off from the congested link and will be realised at a different time or on an alternative route. In this case marginal costs would show a further increase beyond the capacity limit on the considered link. This borderland of marginal social congestion cost functions was not subject to investigation within the UNITE research.

Fig. 1 provides examples of the four types of speed-flow curves listed above and presents the marginal external congestion costs (at the actual traffic level) arising from them. In the example, a motorway with separated carriageways and three lanes per direction without gradients and curves was selected. The travel speeds and congestion costs are depicted for passenger cars at a share of heavy vehicles of 10%. The two types of hyperbolic functions are defined after the definitions given in the EWS manual (FGSV, 1997) for composed functions and in the RAS-W manual (FGSV, 1986) for simple functions. The comparison of the two functions, which are both valid for German traffic planning, gives a illustrative example for the uncertainty associated with the estimation of speed-flow relationships. It is frequently argued, that EWS functions reflect travel behaviour better than the RAS-W functions. The parameters of the two-part linear function is taken from Nash and Sansom (1999). The TRENEN-style functions are just presented as deductive examples .

The comparison of speed-flow curves presented in Fig. 1 reveals, that the expectations of traffic engineers concerning road performance levels differ widely. The difference between the U.K. functions and the German speed-flow curves cannot easily be explained by different patterns of travel behaviour.

The range of marginal external congestion costs arising from these different speed-flow curves is enormous. While the difference for the two German function types up to the capacity limit set by the RAS-W functions is roughly a factor of 3, the congestion costs according to the COBA functions is more than 10 times above the German values.

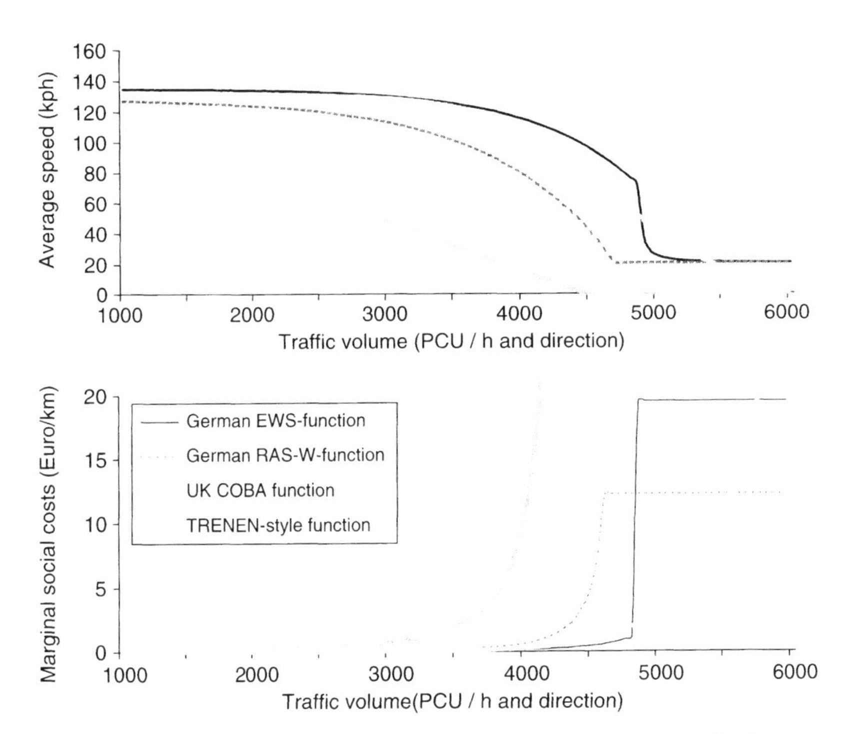

Fig. 1. Travel Speed (I) and Actual External Congestion Costs (II) for Passenger Cars on a Three-Lane Motorway Using Different Speed-Flow Curves. *Source*: IWW.

Moreover, the very different capacity assumptions and the different mathematical formulations of the curves complicate the comparison of the results. Fig. 1 illustrates these difficulties. As it depicts only a relatively small range of the possible marginal cost values the maxima in the lower diagram do not necessarily coincide with the respective minima in the upper graph.

In the UNITE project the German EWS functions have been selected for application to four international inter-urban road corridors, because they represent the most recent available evidence on the inter-dependency of traffic load and travel speed. However, the explanatory power of these types of macroscopic speed-flow relationships for the congestion phenomenon is limited. Although the UNITE research has decomposed the traffic flow into passenger cars and goods vehicles by applying separate speed-flow functions, which mutually take effect on each other, the chosen approach cannot really take account of other vehicle characteristics (e.g. the weight to horsepower ratio) or time-related effects (queuing duration of traffic jams, etc.).

The marginal cost case studies applied to calculate urban road congestion costs made use of the SATURN traffic model developed at the Institute for Transport Studies (ITS) of the University of Leeds. The model computes the performance of traffic flows on the basis of delays at junctions in the urban road network, because junctions are much more critical in urban networks than the road links. The presentation of the underlying junction delay relationships is not possible here as they are an internal model specification, but it must be assumed that their estimation suffers from similar problems to the definition of link-specific speed-flow functions.

In the case of trains and aircraft, scheduled paths are allocated to avoid congestion, but delays still increase as a result of reduced reliability as capacity is approached. Thus, in the rail and air congestion case studies observed data on traffic volumes and delays has been applied. The resulting relationships are of course case-study specific and cannot easily be transferred to other contexts. Moreover, it must be pointed out that the data sets used, especially in the case of rail traffic, were comparably small and aggregated over longer time periods and locations. It is clear that evidence on the mutual interference of users on the basis of such data is poor and thus that results on marginal congestion costs found for rail and air transport need to be considered with care. It is also arguable that the results found by the UNITE case studies on mass transport congestion need to be corrected downwards due to the fact many of the delays are to services of the same operator as caused the delay; thus the interference of trains or aircrafts are anticipated by the operators, and are not fully external.

In economic terms this can be explained as follows: If we presume that the system operator takes the effects of adding another traffic unit to the network into account and follows the goal of optimising the surplus for the entire system, the demand and user cost observations do not describe the average user cost curve for a particular traffic unit. In an extreme case they already describe the social cost function. Consequently, congestion externalities would already be internalised. As this ideal information and behaviour will not be the case in real transport systems, some marginal external congestion costs will remain, particularly where there are many operators and the infrastructure manager does not have social surplus maximising objectives.

The discussion shows the difficulties and uncertainties arising from the application of macroscopic speed-flow curves. A more appropriate, but also much more resource-consuming approach to calculate congestion costs in road traffic as well as in rail or air traffic is the application of traffic simulation models. These could take account of specific vehicle, infrastructure and operation-related characteristics and thus could identify much better the responsibility of certain traffic users for the occurrence of congestion and thus of their marginal external congestion costs.

2.3. Demand Functions and System Specifications

The major driving factors of traffic congestion are the costs of individual users, which increase as more users are on the road. However, there is a feedback as the number of users competing for road space also depends on their average costs. Thus, if user charges are set equal to the actual external congestion costs, demand would fall and with it the external congestion costs. Eventually, neoclassical theory says that the optimal user charge is the marginal external congestion costs at the equilibrium of supply and demand.

In this context the supply curve is represented by the marginal social congestion cost as a function of traffic demand. Here we assume two things: The first assumption is that all demand can be served and that only the costs of serving each unit varies. Although in reality we have queuing effects arising from the given capacity of transport links, this assumption holds true if the time interval considered is long enough (see the above discussion on speed-flow relationships). The second assumption made is that users behave selfishly, e.g. they do not care about the others suffering from congestion when making a travel decision. If this would not be the case, some share of the external congestion costs derived by the Eqs. (1) and (2) would be already internalised and should not be considered when setting congestion externality charges. As to our knowledge no in-depth research on this topic has been conducted, the common assumption on the fully non-altruistic behaviour of users is maintained within the UNITE methodology.

The quantification of the demand function is very difficult in practice as there are various alternative travel decisions of people and companies. The most important ones are:

- *Route choice*: Individual travel drivers may adapt their driving route according to the perceived costs of each alternative available. This means that the charge system on a particular link will have effects on traffic loads, and thus on the marginal external congestion costs at other parts of the network.
- *Departure time choice*: Similar to the route choice effect, shifts in departure time will have effects not only on the demand of a particular network location at the given time, but also on the demand and congestion costs of other time periods.

- *Location choice*: This item subsumes various user reactions in the short- as well as in the long run. In the short run, people might change their travel patterns, e.g. choose different shopping centres or combine work and shopping trips more effectively. Similarly, firms might alter supply or distribution chains. In the medium- to long run people as well as firms may move to save travel costs and/or alter their employer or their production or warehouse patterns, respectively.
- *Overall travel decision*: Individuals may drop unnecessary trips or replace trips by other activities, e.g. Internet shopping, home banking or home shopping, and companies may replace the purchase of input factors by in-house production. Further, a more effective transport organisation can lead to a reduction in trips while the number of passengers or goods transported remains unchanged. In passenger transport this might be done by car pooling, while in freight transport improved loading rates of vehicles can be achieved by improvements in tour planning or the unification or the collaboration of transport companies.

Route shift effects due to changes in transport costs have been widely studied by transport models, in particular in the field of road and rail freight transport. PETS (1999) and STEMM (1998a) study the shift of demand for trans-Alpine freight transport, while STEMM (1998b) reports route shifts for inter-modal trans-Alpine passenger transport. The DESIRE (Design of Inter-Urban Road Pricing Schemes for HGVs in Europe) project (Doll, Kleist, Schade & Schoch, 2003) investigate route shifts of road haulage in different inter-urban road pricing scenarios for Europe and similar studies have been conducted for Germany (Rothengatter & Doll, 2002; IVT/BVU/Kessel & Partner, 2001) and Austria (Herry/IWW/NEA/Snizek, 2002) in order to anticipate the effects of the planned heavy goods vehicles (HGV) charging systems in the two countries. Besides the route shift effect, which is probably the most important determinant of traffic volume on a particular network link, the influence of the remaining driving factors on transport demand is less extensively studied.

The list of driving factors of traffic demand demonstrates, that the consideration of isolated links in a transport network for calculating equilibrium congestion charges is rather difficult. Moreover, it becomes obvious that the explicit determination of a demand function is hardly possible. Thus, UNITE recommends the application of full-scale traffic models, which are based on the microscopic description of individuals and their behaviour, as best practice for the quantification of the congestion phenomenon. As such models were not available for the UNITE case studies on congestion costs, a number of simplifications had to be made.

In case of inter-urban road transport simple constant elasticity demand functions of the form

$$Q(C) = aC^b \tag{3}$$

were applied, where a and b denote model parameters. b is selected such that the overall elasticity of transport demand reflects the modelling experiences concerning network effects, time shifts, etc. on each corridor segment and the scale parameter a was set such that the demand function meets the observed traffic volume. This type of iso-elastic demand function has some very standard mathematical properties, but it can be criticised as the price-elasticity of traffic will vary for different levels of traffic density, for different time segments or for different conditions of the surrounding road network.

The SATURN traffic model, which was applied in the four urban road congestion case studies, in this respect was somewhat closer to reality as network effects have been modelled explicitly. However, other driving factors of demand have also been considered either on a very global basis by using a constant demand elasticity or had been omitted totally.

For the case studies on congestion effects in rail and air traffic it was assumed that the price changes induced by the internalisation of external congestion costs are too small to produce significant demand effects. Thus, actual congestion costs were presented as a proxy for optimal congestion charges.

2.4. Application: The Congestion Cost Case Studies

UNITE has applied 13 case studies on congestion costs covering urban and inter-urban road, rail air and waterborne transport on a large geographical scale. In road and rail transport passenger and freight traffic was considered, while the air case study was restricted to passenger travel and for seaports only freight shipments were considered. In addition, estimates of marginal rail congestion costs found in Sansom, Nash, Mackie, Shires, and Watkiss (2001) for the U.K. were used as benchmark values as there is generally little evidence on the order of magnitude of congestion costs in railway networks.

The results generated by the case studies can be classified into *numerical results* and *analytical results*. Numerical results give the magnitude of user costs and benefits within the case study-specific environment and thus indicate the need for reforming current pricing systems. Analytical results

Table 2. Quantitative Results of the UNITE User Cost Case Studies.

Mode	Area	Range	Notes
Passenger car inter-urban	Paris–Brussels Paris–Munich	0–19 €/100 vkm	Average along corridor for different departure times
HGV inter-urban	Cologne–Milan Mannheim–Duisburg	1–83 €/100 vkm	Average along corridor for different departure times
Car urban	Brussels, Edinburgh, Helsinki, Salzburg	1–34 €/100 vkm	Average in morning peak, varying between cities and geographical locations
Rail passenger services	Switzerland, United Kingdom	0.044–0.30 €/100 pkm	Average across networks varying by time of day and type of service
Air passenger services	Madrid airport	1.8–2.4 €/100 pkm	Average cost values for different years between 1997 and 2000

provide insight into the shape and the determinants of cost functions and thus are important for both the design of optimal pricing strategies and for the transfer of specific numerical results to different spatial, environmental or traffic-related contexts. Table 2 briefly summarises the quantitative results found by the UNITE case studies on congestion and user costs in road, rail, air and sea transport.

All case studies on road congestion costs were carried out using network models. For inter-urban travel the VACLAV traffic model was applied to the four trans-European corridors Paris–Brussels, Paris–Munich, Cologne–Milan and Mannheim–Duisburg, while in urban areas the SATURN model was applied to the cities of Brussels, Edinburgh, Helsinki and Salzburg. Comparing the results it is interesting that the values found for urban peak traffic diverge so much between different cities (5 €-ct. for Helsinki versus 25 €-ct. for Brussels) and that they in some cases are so close to the value found for inter-urban car travel (4 €-ct. as a corridor average). The variability of results on the one hand depends on the road capacity and the traffic load on the corridors, and on the other hand on the time of departure chosen. Not surprising, average marginal congestion costs may be close to zero for over-night journeys.

Congestion costs for rail traffic is found in the range between 1.25 and 41.5 €-ct./train-km according to the case study on Swiss passenger services using an econometric approach and between 14 and 44 €-ct./train-km according to the calculations of railtrack based on a simulation model applied for the U.K. In view of the different methodologies applied by the case studies, it is remarkable that the results are similar in range. Expressed in values per passenger-kilometre these results are only 5% of road congestion costs. Due to the complex time-table structure in rail transport the values will not vary so much between different times of day, but they strongly depend on the network load and on the operation concept in terms of station times, recovery margins and the separation of slow and fast traffic.

Congestion costs in air transport have been studied for Madrid airport using demand delay data from 1996 to 1999 and actual costs of delays for passengers and airlines. Average delay costs of between 1.98 and 68.5 €/flight-km were found, which is much higher per passenger kilometre than the values found for road travel. These delay values are highly dependent on delays at other Spanish and major European airports as well as on en-route traffic density and flight conditions throughout Europe.

3. THE MOHRING EFFECT IN PUBLIC TRANSPORT POLICY ANALYSIS

When turning to the positive effects of additional users of transport systems we also take a step forward in the transport service production process: now we consider the services produced by the transport vehicles of the system, while the capacity of the transport infrastructure to provide services to the vehicles was the focus of attention in the preceding discussion. The present analysis consequently assumes that the negative impacts which the transport vehicles in the system may have on each other in the form of congestion, and accidents are taken into account by appropriate congestion tolls, and accident externality charges on the vehicles.

The "Mohring effect" has its name from a pioneering article by Herbert Mohring in 1972, entitled "Optimization and scale economies in urban bus transportation". As the title suggests, Mohring pointed out that significant economies of scale are prevailing in urban bus transport, which has profound effects both on investment and pricing policy.

At closer scrutiny it turns out that "economies of scale" is a somewhat ambiguous concept in this context, because it is not only a question of the cost effects of firm or plant size (output). It is the density of demand in a given geographical area, or along a given route that matters. In an urban bus transport system, for example, the economies of density of demand takes a number of expressions:

- more bus lines means less access time;
- when bus lines are denser, each line can be straighter to save travel time;
- more buses on each line means less waiting time at bus stops;
- buses should be successively bigger as the density of demand is increasing, which will reduce the bus operator's cost per trip and
- bigger buses can have a higher rate of occupancy with impunity as to queuing time for passengers.

The purpose of the present analysis is to discuss how the Mohring effect could be handled, first when it comes to social marginal cost pricing, and, second, in the determination of the quality of service of public transport. The discussion starts by drawing the attention to the fact that Mohring's finding is a general phenomenon, and a particularly salient feature of the service sector.

3.1. The Generalised Mohring Effect

The transport vehicles of any transport system can be divided into two main groups: vehicles for own account transport and vehicles offering transport services for sale. It is the market for the transport services of the latter group, which is relevant in the present context. Just like all other marketable goods and services there is a production stage, and a distribution stage in the process of meeting the demand for transport services, and it is in the distribution stage, where most of the total "generalised Mohring effect" of an increase in the density of demand appears. Demand density economies in the distribution costs is a general feature throughout the economy, but it is really important only for relatively low-value services.

The illustration of Fig. 2 of three urban areas of the same size shows three different bus transport systems, one consisting of just four buses, another of 16 buses, and a third one of 64 buses. The production stage of the bus transport system is represented by the buses plying the routes of the network. The distribution stage is constituted by the passengers catching the bus, and their getting on and off the bus at the closest bus stop, and the walk to the final destination. It is easily imagined that the more buses there are in the system, implying more lines and more frequent services, the lower the distribution cost per passenger will be, both in terms of walking time, and waiting time at the bus stop.

Suppose that the vehicles in the three Circletown illustrations of Fig. 2 were instead taxicabs. The Mohring effect would be present all the same.

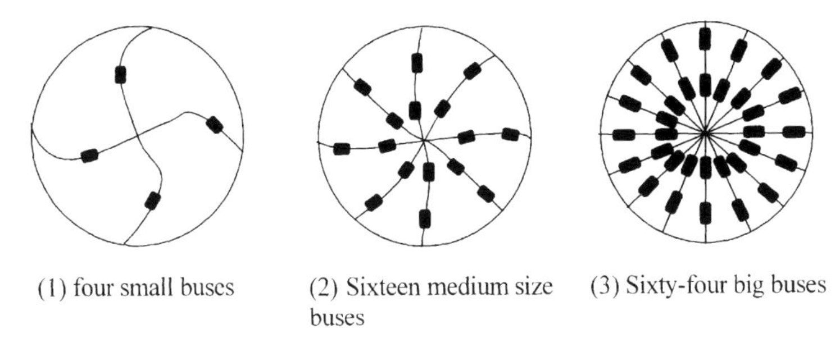

Fig. 2. Buses and Bus Lines in Circletown for Three Different Levels of the Density of Demand. (1) Four small buses, (2) Sixteen medium size buses and (3) Sixty-four big buses.

The main difference is that also the distribution is carried out by the taxis, and the distribution cost is, of course, included in the price of the trip, often in the form of an initial extra charge. However, the greater the density of demand is for taxi transport, the more taxicabs are around, and the lower, on average, the distribution cost will be.[2]

Also outside the transport sector the same phenomenon is omnipresent. Take any goods or service, and it will be found that the greater the density of demand is, the more production plants are required to meet the demand, and the lower the distribution cost per unit will be. It is just that public transport stands out by the fact that the distribution cost is particularly significant relative to the production cost, which to some extent is explained by the additional time dimension of service inaccessibility.

A general expression of the Mohring effect (M) of an additional service user in the service supply system – each user requires one unit of service, let us assume – can consequently be written as the product of the total service volume (B) and the derivative of the average distribution cost with respect to service volume:

$$M = B\,\frac{\mathrm{d}AC_{\mathrm{distr}}}{\mathrm{d}B} \tag{4a}$$

Putting (4a) in an elasticity form, an alternative formulation takes this shape:

$$M = AC_{\mathrm{distr}} \cdot E \tag{4b}$$

where E stands for the elasticity of the average distribution cost with respect to the total service volume. The relative importance of the Mohring effect thus depends, partly on the value of this elasticity, partly on the distribution cost relative to the production cost.

3.2. The Relative Size of the Mohring Effect in Public Transport

In public transport, AC_{distr} mainly corresponds to the user walking and waiting time costs, and the elasticity E is obtained as a weighted average of the walking time, and waiting time elasticities.

In an urban public transport system like that symbolised in the Circletown illustration, the value of the elasticity E can be considered by first observing that the total number of standard-size buses in the system would increase roughly in proportion to the total trip volume. If the frequency of service on each line, and the number of bus lines are increasing in parallel, the value of the walking time elasticity, and waiting time elasticity with respect to trip volume is minus one-half each; the weighted average of these two sub-elasticities would add up to minus one-half, irrespective of in which proportion the required additional buses are used for service frequency, and line density increases.

When it comes to the relative size of AC_{distr}, the comprehensive generalised cost (GC) surveys carried out by the Transport and Road Research Laboratory in England in the 1970s and 1980s are illustrative (Webster et al., 1985; 1977). Table 3 is an adaptation of some results of this work.

Table 3. Cost Structure of Travel to/from Work by Bus in U.K. in 1976 (index Numbers; Total GC for a 2 km Trip = 100).

Cost Component	Trip Length (km)		
	2	5	15
Walking time	28	28	30
Waiting time at bus stop	24	25	27
Riding time	25	33	59
Change of bus	2	3	5
Fare	21	36	85
Frequency delay (= disguised waiting time)	n.a.	n.a.	n.a.
Information cost	n.a.	n.a.	n.a.
Total	100	125	206

Source: Own adaptation of Webster (1977).

As seen in the table, the distribution cost (= walking time, waiting time and change of bus) directly borne by the passengers constitute 54, 45, and 30% of the total GC for 2, 5, and 15 km trips, respectively. It is suggestive that the share of the distribution costs tends to fall with the trip length, and it is generally believed that this tendency will be strengthened, as we move from local urban, to regional and/or inter-urban public transport. One should not be too sure, however, until the items marked as not applicable/available have been considered. ("n.a." in Table 3). In urban public transport it can generally be assumed that the travellers do not bother to use timetables, or memorise the exact departure times, but are content with a general notion of the headway; is it about 10 min or less, they tend to arrive at bus stops and stations unpremeditated. In that case the two last cost components in Table 3, "Frequency delay", and "Information cost" are both zero.

3.3. *MC Pricing of Public Transport and the Mohring Effect*

The question now is, how the sizeable Mohring effect should be handled when it comes to pricing policy with a view to social surplus maximisation? At first sight this question seems to have a straightforward answer, since the Mohring effect as defined above is equal to the user cost component of the price-relevant marginal cost, bearing in mind that the public transport users bear all distribution costs. Optimal price, P^*, consists of these two terms:

$$P^* = MC_{\text{prod}} + B\,\frac{\mathrm{d}AC_{\text{user}}}{\mathrm{d}B} \qquad (5)$$

By assumption, the suffix "user" and the suffix "distr" are interchangeable in the present case. The Mohring effect is a negative term, which acts to reduce the optimal price well below the producer marginal cost (which itself may fall short of the average total producer cost). Before going on to empirical application, however, the old question of which "run" is price-relevant for the marginal cost calculation has to be settled.

If we took a strictly short-run marginal cost approach to scheduled transport services, we might consider that the timetable and vehicle stock would be fixed. In this case the immediate cost to producers of extra traffic is typically negligible, but that to other users is potentially substantial, taking the form of additional crowding of the vehicle, and at peak times of the inability of some users to board the vehicle intended. No Mohring effect is present, but the second term of (2) is a positive crowding/queuing cost.

However, this would be a very short-sighted viewpoint. Producers can often amend timetables at short notice, and acquire new or different vehicles on hire equally quickly. We consider it more appropriate in the case of scheduled transport services to adopt a "medium term" view of marginal cost, in which the vehicle stock can be varied although the infrastructure cannot.

If the timetable and number and size of vehicles can be varied, then there are potentially two ways of adjusting to additional traffic. The first is to use larger vehicles (or longer trains) and the second is to increase the number of vehicles. The first will add to producer costs, but vehicle cost is subject to pronounced economies of size. The second will add rather more to producer costs (indeed, they may well rise proportionately), but this will be offset by benefits to existing users from the increased service frequency and/or density of lines (the Mohring effect). In either case, then, the price-relevant marginal cost is likely to be substantially below the average total cost to the producer.

If the service level were optimal, and there were no significant indivisibilities, then it would not matter which assumption we used in calculating the price-relevant marginal cost – we would get the same answer. Given indivisibilities, we may use whichever assumption seems the most reasonable in the case at hand (for instance, in some cases, it may not be possible to use larger vehicles or longer trains without substantial infrastructure investment; in others, where lack of demand means that high-cost small vehicles are currently used, substituting larger vehicles is the obvious thing to do in response to an increase in demand). But it is always important to consider pricing decisions alongside service-level decisions; calculating marginal cost using a markedly non-optimal service level might give a very distorted answer.

From a practical/empirical point of view, wherever practicable the most appropriate choice would be to consider a capacity addition by substituting larger vehicles, because this would hold the Mohring effect at bay. As we shall see in what follows, the estimation problems as regards the Mohring effect are great, and an additional bonus of this choice is the finding that the price-relevant cost is constant, by and large, along the expansion path.[3]

This important conclusion could be missed by calculating the price-relevant marginal cost on the basis of vehicle number changes only. In a particular least-cost position, P^* would come out the same, as mentioned, no matter which production factor(s) are varied, but keeping the vehicle size constant in the whole trip volume range, would give a false impression of a steadily increasing price-relevant marginal cost along the expansion path.

With reference to expression (5) above for P^*, MC_{prod} would stay constant, while the absolute value of the negative Mohring effect, which according to (4b) can be written as $AC_{\text{distr}} \cdot E$ would decrease since the average distribution cost is steadily falling with increases in the number of vehicles, and the elasticity E is constant. In the extreme it would look like P^* should be negative when the density of demand is very thin. The mistake is that vehicle size should not stay constant along the expansion path, but be increasing with increases in the density of demand.

That P^* should be constant in space, i.e. the same on thin routes, and fat routes alike, is a handy conclusion for public transport system optimisation, but it does not apply to the optimal price structure in time. Peak-load pricing would be very useful, in particular, on long-distance routes, but this aspect of pricing policy is not taken up here (for a further discussion, see e.g. Jansson, 2001a).

A second conclusion is that by the choice of the particular "medium run" marginal cost variant suggested, it is not necessary to estimate the Mohring effect for the purpose of pricing policy. However, pricing is only one aspect of a social-surplus maximising public transport policy. To determine the optimal level of service, the value of the Mohring effect is essential.

3.4. The Headway Cost in the Whole Range of Service Frequency

The attractive simplicity of the high-frequency case discussed above is not a general characteristic. Mean waiting time at stops and stations equal to half the headway is a convenient assumption for a substantial part of urban public transport, but is just an approximation, which will be less and less accurate as the headway is increasing. The relationships between mean waiting time and headway of the diagrams in Fig. 3 is an illustration of the basic fact that travellers increasingly take the trouble to consult a timetable in advance when headways exceed 10 min. They no longer arrive at stops and stations completely at random, but a larger and larger proportion of the travellers adjust their expected arrival at the stop or station where they will board the bus or train to a couple of minutes (safety margin) before scheduled departure time.

Timetables which can be trusted are obviously a good thing, which limits the deliberate waiting time at stops and stations to 4 or 5 min, but what is bad with infrequent public transport services is that the arrival time at the destination stop or station will, on average, be more and more out of line with the travellers' wishes.

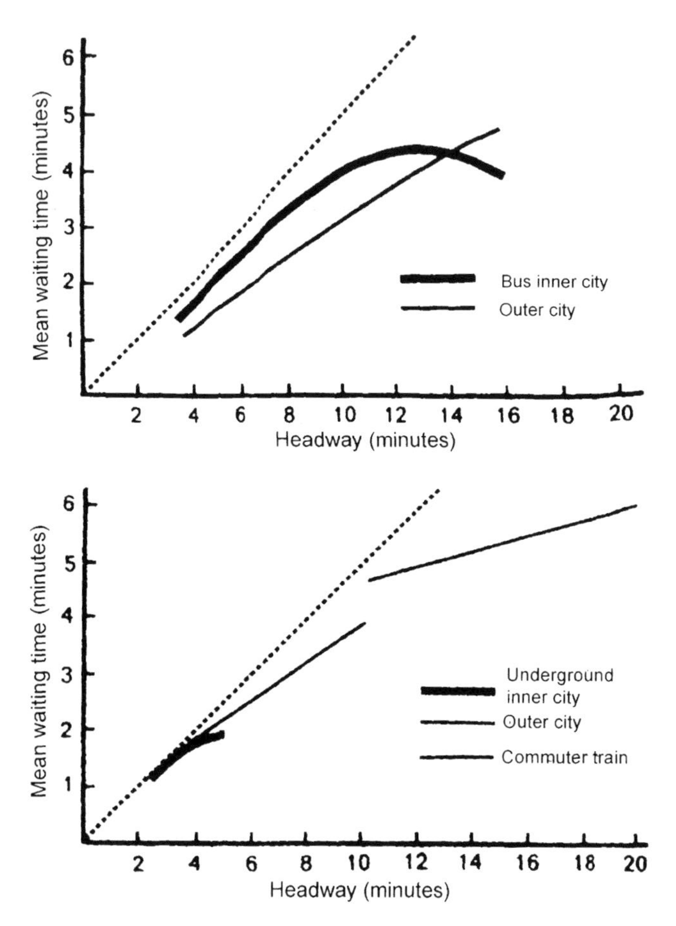

Fig. 3. Mean Waiting Time for Buses and Trains in Stockholm. *Source*: TU:71: Trafikundersökningar i Stockholmsregionen hösten 1971, resultatrapport nr 1. Stockholm's Läns Landsting, Trafiknämnden.

3.4.1. Conceptual Discussion: Frequency Delay

This failure of scheduled public transport to provide arrival times in accordance with travellers' wishes was called "frequency delay" in a pioneering paper by John Panzar looking at airline services in U.S.A. (Pangar, 1979), and is in Scandinavian usage often characterised as "disguised waiting time". It is not a matter of easily observable waiting time at bus stops or railway stations, but is an inconvenience, which can take many

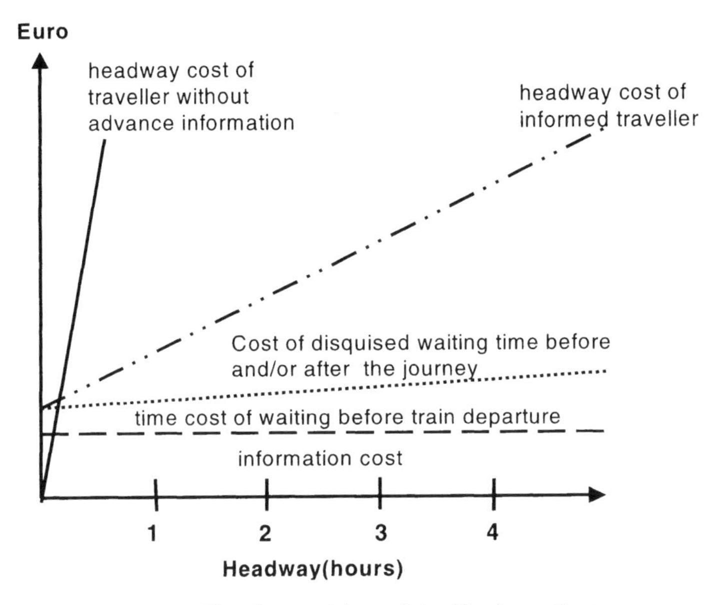

Fig. 4. The Composition of the Headway Cost.

different forms. In some cases it is not a question of waiting time at all, but the drawback and/or embarrassment of being late.

The identification, and quantification difficulties concerning the frequency delay must not be taken as a pretext for ignoring it, or underrating its importance. It can be argued that the main reason why public transport services are simply lacking in most non-urban relations, which is usually put into words like "the demand base is too thin for a viable service", is more pointedly explained by a prohibitively high cost of the frequency delay that would be incurred.

The problem is that the exact value of this cost is difficult to come to grips with. As a starter in the quest for this elusive cost, the diagram of Fig. 4 is offered in order to sort out the conceptual confusion. The basic assumption is that the main cost driver is the headway (h), i.e. the time distance between the vehicles on the route concerned.

Generally speaking, there are two categories of public transport users who should be distinguished: those who do not bother to consult a timetable, and those who do.

The headway cost per trip of the former category can be confined to the actual waiting time cost at stops and stations represented in the diagram by the steep curve right from the origin. Its steepness indicates that the cost per minute of actual waiting time would be considerably more onerous than "disguised waiting time" taking place at home, or somewhere else at the origin and/or destination.

Those who take the trouble of informing themselves beforehand of the schedule can be assumed to incur an initial "information cost". That there is a sacrifice involved in getting and assimilating the timetable information is clear from the fact that the travellers trade in expected waiting time for not having to obtain the information. This trade-off ceases fairly soon after the headway has passed a quarter of an hour.

For very high service frequencies almost everybody belong to the former category, and for low service frequencies, corresponding to headways beyond a quarter of an hour, almost everybody is informed well in advance. This does, of course, not mean that the headway cost stops rising with further lengthening of the headway. The inconvenience of the frequency delay or "disguised waiting time" is rising indefinitely; the question is, at which rate?

Besides the theoretical underpinning of the interpretation of the available empirical evidence afforded, the point of the fundamental dichotomy is to draw the attention to different possibilities to reduce the headway cost. For example, the steeply rising cost function of travellers arriving to a bus stop, or metro station without advance information can be rotated downwards by actual on-line information of the time of next departure; waiting without knowledge of how long the waiting time will be can be very toilsome. As regards the other category of travellers, a parallel downward shift of the headway cost function could be obtained by other information improvements.

3.4.2. The Empirical Headway Cost Function

The final headway cost function, which comes out of empirical studies, is basically a combination of the headway cost function of travellers without advance information, and the headway cost function of informed travellers. This combined function (which is not depicted in the diagram of Fig. 4 illustrating just the basic principle) depends obviously on the composition of the public transport users with respect to their information behaviour, which most likely varies between different services, and which anyway is quite difficult to observe.

The main reason why the empirical evidence, exemplified by Tables 4 and 5, shows a markedly decreasing marginal headway cost with respect to the headway is likely to be that an increasing part of the travellers change

Table 4. Values of Headway Reduction for Private Trips According to the Swedish CBA Manual used for Transport Investment Planning, Euro per hour.

Headway Range, min	Short-Distance Trip (<50 km)	Long-Distance Trip (>50 km)
<10 min	6.5	3
11–30 min	2.0	3
31–60 min	1.8	3
61–120 min	1.1	1.5
>120 min	0.6	0.7

Source: SIKA (1999).

Table 5. Values of Headway Reduction for Business Trips According to the Swedish CBA Manual used for Transport Investment Planning, Euro per hour.

Headway Range, min	Train	Air
<60 min	10.8	12.9
61–120 min	7.5	10.8
>120 min	6.5	8.6

Source: SIKA (1999).

information behaviour, and take the trouble to consult a timetable as the headway increases beyond 10 min. Another reason may be that some more marked non-linearities are involved as regards the two alternative headway functions in Fig. 4, each representing a pure behavioural assumption. However, in the range of very wide headways, the existing values are rather uncertain. It is even doubtful whether the very approach, which treats the frequency delay as a GC component alongside travel in-vehicle time, and actual waiting time is the right way.

As seen in Table 4 the headway cost per additional unit of time in the upper range where the headway exceeds 2 hours is only a tenth of the headway cost in the range below 10 min for short-distance trips. For long-distance trips the corresponding values differ by a factor of 5.

The values representing mainly actual waiting time at stops and stations are probably more reliable than the values at the other end of the headway range. In Sweden, we stick to the time-honoured convention in urban transport planning of doubling the VOT on the bus to represent the value of waiting time at the bus stop. In Britain two authoritative sources narrow down the likely true value to a range consistent with this convention. One recommends a factor of 1.6 rather than 2.0 for walking and waiting time as a result of a meta-analysis including no less than 43 studies where values of walking time, and 13 studies where values of waiting time are reported (Wardman, 2001), and the other recommends "that waiting time justifies a weight on average of two and a half rather than two relative to in-vehicle time" (Mackie et al., 2003).

3.4.3. An Alternative Approach

The headway cost figures of Tables 4 and 5 are used in practice mainly by the Swedish Railtrack ("Banverket") for rail investment Cost–Benefit Analysis (CBA). The train operators in Sweden, of which the Swedish Railways (SJ) still has a dominant position, do not use CBA, but Cost–Revenue Analysis like ordinary profit-maximising business firms. This was also true of SJ before deregulation, although SJ was perhaps better described as a loss-minimiser before its recent financial restructuring.

It is interesting to note that a profit-maximiser should out of self-interest choose the same quality of service as a social surplus-maximiser for each given level of output (trip volume).[4] Therefore an alternative method of estimating the headway cost to the main approach of travel demand analysis is to explore the trade-off that train operators make between increased costs for themselves of higher frequency of service, and lower headway costs for their passengers.

For this purpose a cost model of a railway line was developed in a case study under the umbrella of the UNITE project, where both producer and user costs are included (Ericsson & Jansson, 2001). All factor prices were known to us except the headway cost per trip, which moreover could not be assumed to be independent of the headway (h), but should be thought of as a function $C(h)$. The idea was to minimise the total costs and solve for h from the minimum cost condition and compare the result with the pattern of the input of engines, and carriages provided by SJ on different rail lines served by flexible-formation trains. This pattern was surveyed, and analysed in a parallel, empirical case study (Jansson, 2001).

The latter was carried out as a cross-section study of the number of train departures per day, and the total number of passengers per day on 32 lines of widely different distances in the Swedish rail network. Its main result is that, if frequency of service is plotted against the square root of the number of passengers per line kilometre (Fig. 5), a linear function with the origin as starting point fits the observations very well: the adjusted $R^2 = 0.955$, and the t-value $= 27$.

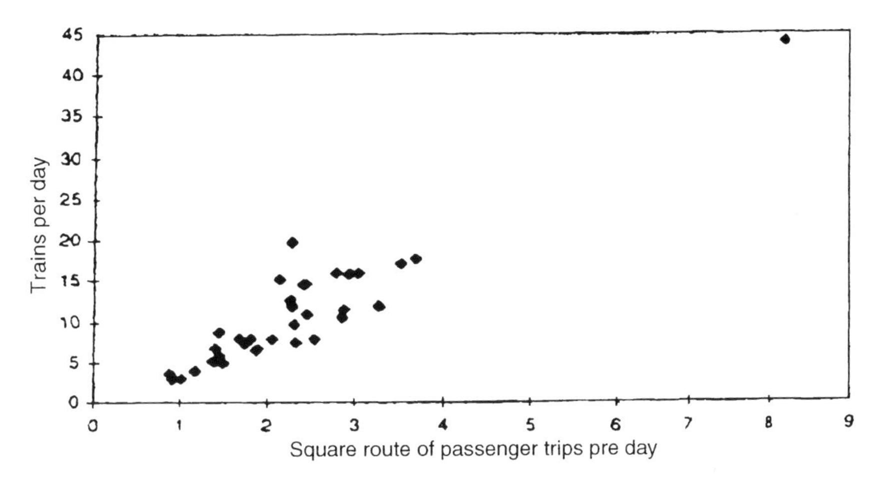

Fig. 5. Frequency of Service Versus the Square Root of Passenger Trips/Day. *Source*: Persson (1999).

The railway line optimisation model shows that a corresponding "square root law" applies in case the headway cost per trip is proportional to the headway, i.e. constant per unit of the time interval between train departures. Therefore, the first conclusion is that the SJ managers who were responsible for the flexible formation of train services in the 1990s believed that the headway cost function takes this shape.

It can be noted that the "square root law", i.e. proportionality between frequency of service, and the square root of the density of demand, applies independent of the actual value of the headway cost per unit of time, as long as a linear shape can be assumed. However, by comparing the proportionality constant found in the regression analysis with the result of the optimisation model, it is concluded that railway managers behave as if they believe the headway cost per hour to be about 2 €, on average. (This figure is to be interpreted as an average of private and business trip values).

This value can be compared to the values used by the Swedish Railtrack (Banverket, 2000) in order to take the Mohring effect into account in CBA of rail track investments given in Tables 4 and 5. The headway cost function assumed there is not linear, but gives costs per unit of time which are tapering off with increases in the headway: for example, below half an hour the value is 2 €/hour (that is 2/60 €/min) for private trips, and above 2 hours' headway, the value is only 6 €/hour. A rough average value, including both private and business trips on different lines with widely different demand density, would be closer to 3 € than to 2 €/hour, so it seems that the belief of SJ was a little more conservative as regards the value to passengers of higher frequency of service than the belief of Banverket (the Swedish Railtrack).

4. CONCLUSION

Congestion costs depend on the value of time, speed-flow relationships and demand function. They have been widely studied for road, but less so for rail, air and maritime. Moreover it is less clear to what extent they should always be regarded as truly external for scheduled modes. For example, while it is fairly clear that the costs of congestion-related delay imposed by one train operator's trains on another train operator are external, it is debatable whether congestion-related delays caused by the presence of trains of the same train operating company should be regarded as price-relevant or as already internalised.

The costs and benefits of user interference is in the first place driven by traffic demand and therefore congestion costs vary substantially from case to case. Nevertheless we can learn from the case studies, that congestion is highly important for road and air traffic but less important for rail.

Mohring benefits are usually thought of as applying to high-frequency urban transport, where waiting time is reduced by more frequent service. However they may also apply to lower frequency scheduled services, when an increase in frequency permits a closer matching of desired and actual schedules, although as a proportion of total cost their significance here will be smaller.

When existing frequencies are roughly optimal and it is feasible to operate larger vehicles, the marginal cost of increasing vehicle size is the easiest concept to estimate; in this situation, valuation of the Mohring effect is not needed for pricing policy. It is only where constraints prevent the operation of larger vehicles that the size of the Mohring effect needs to be known. Evidence is provided that the Swedish rail operator adjusts frequencies in proportion to the square root of demand.

NOTES

1. Claus Doll was responsible for Section 22 and Jan Owen Jansson for Section 33.
2. See further Jansson (2005).
3. This is not demonstrated here, but is shown in Jansson (1984).
4. For a carefully prepared analysis of this proposition, see Sonesson (2005).

REFERENCES

Alger, S., Hugosson, B., & Lindqvist-Dillén, J. (1995). *1994 Års Tidsvärdesstudie, final report and results*. Study commissioned by the Swedish Institute for Transport and Communikation Analysis (SIKA). Transek AB, Solna.

Banverket. (2000). Beräkningshandledning – Hjälpmedel för samhällsekonomiska bedömningar inom järnvägssektorn. BVH 706, Borlänge.

Blonk, W. (1999). *Final report on estimating transport Costs*. Final Report of the High Level Group on Transport Infrastructure Charging of the European Commission, DG-VII, Brussels.

Department for Transport (2002). Economic assessment of road schemes. *The COBA manual* (Vol. 13, Section 1). London, May.

de Jony, G. (1996). Freight and coach VOT studies. Paper presented at the Value of Time Seminar organised by the PTRC, 29–30, October 1996, Wokingham, UK.

Doll, C., Kleist, L., Schade, W., & Schoch, M. (2003). Input analysis and assessment. DESIRE (*Design of Inter-Urban Road Pricing Schemes*), *Deliverable D4*. Project funded by the European Commission, DG-TREN, Karlsruhe.

Ericsson, R., & Jansson, J. O. (2001). Price-relevant marginal cost of scheduled public transport: Case study of Swedish interurban rail passenger transport with emphasis on the supplier's marginal cost. UNITE Wp6, EKI, Linköping University.

FGSV. (1986). *Richtlinien für die Anlage von Straßen (RAS), Teil: Wirtschaftlichkeitsuntersuchungen (RAS-W)*. Research Society for Road and Transportation Science, Cologne.

FGSV. (1997). *Empfehlungen für Wirtschaftlichkeitsstudien an Strassen (EWS)*. Research Society for Road and Transportation Science, Cologne.

Gunn, H. F., & Rohr, C. (1996). The 1994 U.K. VOT study. Paper presented at the Value of Time Seminar organised by the PTRC, 29–30, Wokingham, UK.

Gunn, H. F., Tuinenga, J. G., Cheung, Y. H. F., & Kleijn, H. J. (1999). Value of Dutch travel time savings in 1997. Transport Modelling/Assessment. In: H. Meersman, E. Van de Voorde & W. Winkelmans (Eds), Proceedings of the 8th world conference on transport research (Vol. 3, pp. 513–526). Amsterdam: Pergamon.

Herry/IWW/NEA/Snizek. (2002). *Tarifrechnung ASFINAG. Teilbericht Verlagerung*. Vienna: Study on behalf of the Motorway and Express Route Financing Company (ASFINAG).

IVT/BVU/Kessel & Partner. (2001). *Verlagerungen durch Einführung einer Lkw-Maut*. Berlin: Study on behalf of the Federal Ministry for Transport, Building and Housing.

Jansson, J. O. (1984). *Transport system optimization and pricing*. New York: Wiley.

Jansson, J. O. (2001). *The Mohring effect in interurban rail transport*. UNITE Wp7, EKI: Linköping University.

Jansson, J. O. (2001a). *Efficient modal split*. Thredbo 7, Molde, 25–28. June.

Jansson, J. O. (2005). *The economics of services: Development and policy*. Cheltenham, UK: Edward Elgar.

Link, H., & Maibach, M. (1999). *Calculating transport infrastructure costs*. Final Report of the Experts Advisors to the High Level Group on Transport Infrastructure Charging of the European Commission, DG-VII. Berlin, Zurich.

Mackie, P.J., Wardman, M., Fowkes, A.S., Whelan, G., & Nellthorp, J. (2003). *Values of travel time savings in the UK*. Report to Department of Transport, ITS, University of Leeds.

Mohring, H. (1972). Optimization and scale economies in urban bus transportation. *American Economic Review*, *62*(4), 591–604.

Nash, C., & Sansom, T. (1999). *Calculating transport congestion and scarcity costs*. Final Report of the Expert Advisors on the High Level Group on Infrastructure Charging (Working Group 2).

Nellthorp, J., Mackie, P. J., & Bristow, A. L. (1998). *Measurement and valuation of the impacts of transport initiatives*. Deliverable D9, (Restricted), EUNET Project – Socio Economic and Spatial Impacts of Transport (Contract: ST-96-SC.037), Institute for Transport Studies, University of Leeds.

Panzar, J. (1979). Economics and welfare in unregulated airline markets. *American Economic Review*, May

Persson, J. (1999). *SJs implicita värdering av turgleshetskostnaden*. Working paper, EKI, Linköping University.

PETS. (1997). *Internalisation of externalities*. Deliverable 7 of the research project PETS (Pricing European Transport Systems), financed by the 4th framework programme of the European Commission. Project-co-ordinator: Institute for Transport Studies, University of Leeds.

PETS. (1999). *Case study: Trans-alpine freight transport*. Deliverable 11 of the project PETS (Pricing European Transport System) funded by the 4th RTD-programme of the European Commission. Project Co-ordinator: Institute for Transport Studies, Leeds.

Proost, S., & van Dender, K. (1999). *Final report of the TRENEN-II-STRAN project*. Project financed, by the European Commission, Brussels.

Rothengatter, W., & Doll, C. (2002). Design of a user charge for heavy goods vehicles on German motorways considering the objectives of efficiency, fairness and environmental protection. *IATSS Research*, *26*(2) 28 ff.

Sansom, T., Nash, C., Mackie, P., Shires, J., & Watkiss, P. (2001). *Surface transport costs and charges Great Britain 1998*. Study commissioned by the Department of Transport, Environment and the Regions. London, Leeds.

SIKA. (1999). Översyn av samhällsekonomiska kalkylprinciper och kalkylvärden på transportområdet. ASEK, Rapport 1999:6.

Sonesson, T. (2005). *Optimal system of subsidisation for local public transport*. Vinnova/EKI: Linköping University.

STEMM. (1998a). *Trans-alpine passenger transport*. Deliverable 6A of the project STEMM (Strategic Multi-Modal Modelling) financed by the 4th RTD framework programme of the European Commission. Project co-ordinator: Baxter Eadie Ltd, London.

STEMM. (1998b). *Trans-alpine freight transport*. Deliverable 6B of the project STEMM (Strategic Multi-Modal Modelling) financed by the 4th RTD framework programme of the European Commission. Project co-ordinator: Baxter Eadie Ltd, London.

Wardman, M. (2001). A review of British evidence on time and service quality valuation. *Transportation Research*, *37E*(2/3), 107–128.

Webster, V. (1977). *Urban passenger transport, some trends and prospects*. Crowthorne, UK: Transport and Road Research Laboratory.

Webster, V., Bly, P. H., Johnston, R. H., Paulley, N., & Dasgupta, M. (1985). *Changing patterns of urban travel*. Crowthorne, UK: Transport and Road Research laboratory.

ACCIDENTS

Gunnar Lindberg

1. INTRODUCTION

Accidents are a severe problem in all modes of transport. The consequence of a traffic accident can be horrendous, and people indicate a high willingness-to-pay (WTP) to reduce the risk of being the victim of a traffic accident. A part of the traffic accident problem can be explained by the fact that the user, in his or her decision, does not consider all costs related to an accident – a part of the accident cost is external to the user.

An accident is an outcome of a number of unlucky circumstances of which some can be influenced by the driver such as his or her number of trips, driving behaviour and choice of vehicle type. If some of the accident costs are external, decisions regarding each of these factors will be erroneous. The speed will be too high, the vehicles will be too heavy and the number of trips will be too many. In the following, we focus on the last decision, the number of trips.

Let us assume that the marginal external accident cost is estimated and introduced as a distance-based charge. What should be included in the charge? *Firstly*, we need to find the risk and the related cost of the charged users' accidents. *Secondly*, we have the question of whether we need to remind the users of their own accident costs. If not, a large part of the accident cost is already internalised in the users' decisions, and it is not necessary to remind them of this cost once again. It is suggested here that the user is aware of the risk and "cost" of being a victim. *Thirdly*, in which way the risk and cost will change with an additional trip or an additional

Measuring the Marginal Social Cost of Transport
Research in Transportation Economics, Volume 14, 155–183
Copyright © 2005 by Elsevier Ltd.
All rights of reproduction in any form reserved
ISSN: 0739-8859/doi:10.1016/S0739-8859(05)14006-2

kilometre driven will have a strong impact on the external marginal accident cost.

The remainder of the chapter is organised in the following way. Section 2 discusses the important issues of valuation of accidents while Section 3 develops the theoretical frame of the external marginal accident cost. The following Section 4 deals with two detailed case studies, railway level crossing accidents and accidents where heavy goods vehicles (HGVs) are involved. Section 5 presents some remaining issues such as risk-avoiding behaviour and the role of liability.

2. VALUATION OF ACCIDENTS

A traffic accident with casualties is both an invaluable human tragedy and a huge loss of economic resources. While some of the costs are visible, others are more difficult to observe. The medical treatment of an accident will for some patients continue over many years ahead, in the worst cases over their whole lifetime. Both fatalities and the most severe injuries will deprive the economy of labour resources for the whole expected lifetime of these victims. But most importantly, the human tragedy comes with a cost above the loss of economic resources. People value their safety for more subtle reasons than their lost production capacity.

The valuation of an accident can be divided into *direct economic costs*, *indirect economic costs* and a *value of safety per se*. The direct cost is observable as expenditure today or in the future. This includes medical and rehabilitation costs, legal cost, emergency services and property damage cost. The indirect cost is the lost production capacity to the economy that results from premature death or reduced working capability due to the accident.

However, these two components do not reflect the well-being of people. People are willing to pay large amounts to reduce the probability of premature death irrespective of their production capacity. The WTP estimates the amount of money people are willing to forgo to obtain a reduction in the risk of death. Two basic methods can be used to estimate this WTP,[1] *revealed preference* or *stated preference*. The former is based on actual market transactions by the individuals. The most frequent technique is wage-risk studies, which estimate the wage premium associated with the fatality risk at work (see Viscusi & Aldy, 2003). The main disadvantage in revealed preference studies is the difficulty to find a distinctive traffic safety product on the market. In its place, stated preference methods have been the preferred method to elicit a value of safety *per se*. A hypothetical market situation is created, which people are asked to value. A typical study would describe the safety situation on roads and then ask for the WTP of a private product or a public programme that increases the safety by, say, 10%.

The marginal rate of substitution of wealth for probability of death is transferred to the value of statistical life (VOSL). Assume that the WTP is expressed for a small change in the probability (dz) for one individual. Sum these small changes over n individuals such that $n\,dz = 1$; i.e. in a group of size n one statistical death will be avoided. The VOSL may then be expressed as the sum of the WTP over all n individuals.

$$\text{VOSL} = \text{WTP}\,n = \text{WTP}/dz \tag{1}$$

Recently, serious questions on the reliability of stated preference results have been raised. Serious problems with hypothetic bias and embedding effects have been found. When people are confronted with a hypothetical question the answer does not always reflect their actual behaviour, and the hypothetical WTP often exceeds the actual WTP. This could be due to the lack of a budget constraint but also that the question introduces a product that is not sold on the market; the respondent has no experience to trade with this product. Embedding effects can be owing to what is known as the "warm glow effect", that is, the responses "reflect the WTP for a moral satisfaction of contributing to public goods, not the economic value of these goods" (Kahneman & Knetch, 1992 p. 57). Another possible explanation is that it can be difficult for the respondents to understand small changes in small probabilities; this can be called scale bias. The result is that respondents report the same WTP for a larger safety improvement as for a smaller improvement. If the responses are only weakly dependent on the magnitude of the risk reduction, almost any VOSL can be derived from the studies. In Beattie et al. (1998), the VOSL is estimated at 16.4 million € for the risk reduction 1/100,000 and 7.7 million € for a three times higher risk reduction.

Throughout the world empirical estimates of VOSL diametrically differ between different studies, ranging from a value of less than 200,000 to 30 million U.S. dollars (Blaeij, 2003). Making meta-analysis of this material is difficult, and it is important to focus on the reliability studies. Carthy et al. (1999) use the contingent valuation method to estimate a WTP for less severe outcomes and a risk/risk analysis to link this WTP to fatality. The authors suggest the range 1.0–1.6 million £. This value includes the cost of net lost production, medical and ambulance costs (c in Table 1). If these costs are excluded, a recommended value will be 1.5 million € (a in Table 1).

Table 1. Indicative Accident Values and its Components (million €).

	Fatality	Severe Injury	Slight Injury
VOSL (*a*) based on WTP	1.50 M€	0.20 M€	0.015 M€
Additional accident cost for the rest of the society (*c*)	8%	20%	20%
Total cost I ($a+c$)	1.62 M€	0.23 M€	0.02 M€
Additional WTP based on relatives and friends (*b*)	40%	40%	40%
Total cost II ($a+b+c$)	2.22 M€	0.31 M€	0.02 M€

For less severe consequences, ECMT (1998) suggests that the value for severe injuries is 13% and for light injuries 1% of VOSL of fatalities. Newer evidences do not suggest that these values need to be updated. However, it would be an important improvement if the group severe injury was split into two groups, permanent and temporary injuries, which shows a huge difference in WTP (Persson, Hjalte, Nilsson, & Norinder, 2000).

When adding VOSL to the estimate of accident cost, double accounting may occur in relation to gross production losses. It is often assumed that VOSL includes the value of lost consumption of the deceased person. This is also included in the gross lost production. Two possibilities exist; to reduce the deceased persons' consumption from the gross lost production and express it as *net lost production* or to reduce VOSL with the amount of lost consumption,which results in a so-called *human value*.

So far we have discussed safety as a purely selfish problem. The affected individual may have relatives outside the household, and friends who care about his exposure to risk, and consequently have a WTP for his risk reduction. Although only very few studies are aimed at estimating relatives' and friends' valuation, a value of around 40% of the selfish value seems to be justified.[2] Based on the argument put forward by Hochman and Rodgers (1969) and Bergstrom (1982), no value reflecting relatives' and friends' valuation should be included for public investments under the assumption of *pure altruism.* The intuition is that a pure altruist would care both for other people's safety, and the lost well-being related to the cost they have to pay to have more safety. When it comes to accident externality charges, this valuation problem is less delicate. When a car user increases the risk for a pedestrian to be killed in an accident, the welfare of the pedestrian's relatives and friends will also be reduced, which should add to the accident externality charge on the car user, since it would not affect the financial situation of the pedestrian. On the other hand, when it comes to the question if the prospective victim should pay a charge related to an increase in his own accident risk, his relatives' and friends' valuation of reduced consumption and improved safety for him may cancel out in the same manner as in the public investment case.

In conclusion, the dominant element of the accident cost is the VOSL (a). A part of the direct cost and indirect cost is included in the VOSL and should not be added separately. The cost to add consists of the cost to the rest of society (c). This is both costs for hospital care paid by the general tax system and net lost production, which would have gained the society at large. Table 1 presents some evidence of the magnitude of these elements from Sweden. However, the ratio between the components differs between different social security systems. Finally, we indicate the magnitude of relatives' and friends' valuation (b).

3. MARGINAL COST OF ACCIDENTS

Let us assume that a car driver or public transport operator is (i) aware of the accident risk related to trips and (ii) will bear all costs in the case of an accident. Under these circumstances, they will be aware of the expected accident cost that will arise due to each trip decision. This cost, together with vehicle operating costs, time costs and charges for other external effects, will be weighted against the expected revenue or benefit of the trip. Under these circumstances the trip decision and other safety-related decisions, such as driver behaviour or choice of vehicle type, will be socio-economic efficient. If the driver does not bear all accident costs, an accident externality will emerge.

A theory of external accident cost has been discussed in Vickrey (1968), Newbery (1988), Jones-Lee (1990) and Jansson (1994). Vickrey and Newbery concentrate exclusively on the relationship between traffic volume and accidents in homogeneous car traffic. Jones-Lee (1990) allows for both car and unprotected road users but assumes the risk to be independent of traffic flow. Jansson (1994) consolidates these two approaches in separate models for homogeneous (only cars) and heterogeneous traffic (cars and other users). In Lindberg (2001), a more general theory covering all cases based on Jansson's approach was presented.

These discussions focus on the externality by driven distance, but it also exists as an externality in driver behaviour (speed choice) or vehicle choice (sport-utility vehicle (SUV) versus small compact cars). Nevertheless, we will continue here to focus on driven distance. We assume that the charge shall be paid *ex ante*, i.e. before the trip has started and before an accident occurs. *This charge should be based on the expected accident cost the driver/ operator does not already bear, and that arises as a consequence of his or her trip decision.* The external marginal accident cost as a basis for this charge will depend on the:

I. *magnitude of the risk,*
II. *associated accident cost,*
III. *the accident cost the decision maker does not bear, and*
IV. *how this cost is influenced by his or her trip decision.*

3.1. A Formal Derivation of the Marginal Cost

The total annual cost of accidents (TC), where vehicle type j has been involved, can be written as Eq. (2) where A is the number of accidents and $(a+b+c)$ the cost components discussed in Section 2. With involved we mean that the vehicle has been one of the parts in the accident, irrespective of who was hurt or who was at fault. The risk (r) of category j to be involved in an accident (3) may be affected by an increase in the volume of traffic of category j (Q). This effect is expressed as a risk-elasticity (E), Eq. (4). The marginal cost with respect to the traffic volume for a vehicle of category j can be written as Eq. (5). We derive the external marginal cost as Eq. (6), where PMC is the private marginal cost already internalised. If we introduce θ as the share of the accident cost per collision that falls on category j (7), the external marginal accident cost can conveniently be expressed as (8).

$$\mathrm{TC}_j = A(a+b+c) = rQ(a+b+c) \tag{2}$$

$$r = \frac{A}{Q} \tag{3}$$

$$E = \frac{\partial \mathrm{r}}{\partial \mathrm{Q}}\frac{Q}{r} \tag{4}$$

$$\mathrm{MC}_j = \frac{\partial A}{\partial Q}(a+b+c) = r(E+1)(a+b+c) \tag{5}$$

$$\mathrm{MC}_j^e = MC - \mathrm{PMC}_j \tag{6}$$

$$\theta = \frac{\mathrm{PMC}}{r(a+b)} \tag{7}$$

$$\mathrm{MC}_j^e = r(a+b)[1-\theta+E] + rc(1+E) \tag{8}$$

We expect the external marginal cost to be high if:

- *the accident risk r is high;*
- *the cost per accident is high* $(a+b+c)$;
- *most of the costs fall on other user groups* $(\theta \approx 0)$;
- *the risk increases when the traffic increases* $(E>0)$;
- *or a large part of the accident cost is paid by the society at large* (c).

In addition to the costing of accidents discussed in the previous section, it is thus necessary to understand the division between internal and external costs, and the relationship between risk and number of trips.

3.2. Internal and External Accident Costs

The question of internal and external accident costs can be broken into two parts; (i) do users consider their own risks and (ii) do they consider the risk of others? The straightforward assumption is *yes* on the first question and *no* on the second question. When VOSL is estimated, users are asked about their trade-off between accident risk and money. The reply is used to derive VOSL. We thus believe that they can value changes in hypothetical risk. If they, in a real situation, also understand and value risk changes the VOSL will be internal. If they do not understand the risk related to their decision, we have an information failure. It has often been shown that individuals overestimate small risks and underestimate large risks – "in general, rare causes of death [are] overestimated and common causes of death [are] underestimated".[3] This has a corollary that they underestimate risk changes.[4] However, recent results suggest that with a more detailed definition of the risk, the timing of the risk[5] or using risk defined for relevant age groups,[6] the difference between actual and perceived risk diminishes or disappears. But, exactly how individuals assess the *marginal* risk related to a change in driven distance or a new trip is unclear. Here have based our analysis on the assumption that they understand the risk change. This is not a trivial assumption.[7]

3.3. The Risk Elasticity

It is well known that when the traffic volume increases on a road, the speed goes down and the average travel time increases. But what about the accident risk? As the number of vehicles increases, the number of accidents will most probably increase; we have not seen any evidence on the opposite effect. However, exactly how the number of accidents increases is important;

will the number of accidents increase in proportion to the increase in traffic volume, or will the increase be progressive or regressive? If the number of accidents increases in proportion to the traffic volume, the risk, i.e. the number of accidents per vehicle or vehicle kilometre, will be constant; the risk elasticity (E) will take the value nil. If the increase is regressive, the accident risk will decline and the elasticity will be negative. This means that an additional user reduces the risk for an accident for all other users. Finally, if the number of accidents increases progressively the risk will increase. An additional vehicle will impose an increased threat to all other vehicles and the external effect will be larger, the elasticity will be positive.

As the number of vehicles increases, the number of possible interactions increases with the square. This suggests that the risk should increase with traffic volume. Dickerson, Peirson, and Vickerman (2000) find that the accident elasticity varies significantly with the traffic flow. They argue that the elasticity is close to zero for low-to-moderate traffic flows, while it increases substantially at high traffic flows. Vitaliano and Held (1991) show in their estimation that the relationship between accidents and flows is nearly proportional and thus the risk elasticity is close to zero. This is also found by Fridstrøm, Ifver, Ingebrigtsen, Kulmala, and Thomsen (1995). However, our results reported in the next section suggest that the risk decreases with traffic volume. In an overview of six international studies, Chambron (2000) finds a less than proportional increase in injury and fatal accidents. This has also been found by Hauer and Bamfo (1997), and a majority of the results reviewed in Ardekani, Hauer, and Jarnei (1997). The reason for this decline in risk is discussed in Section 5.

3.4. A General Example

The matrix given in Table 2 depicts accident cost in relation to road transport accidents. The column to the left shows the victims by different types of users involved in road accidents, and the row on the top shows the different types of accidents in which the users may be involved. To create such a matrix rather specific demands has to be put on the accident-recording system, but it is necessary if we want to estimate the external marginal cost.

As an example, we use information from Sweden. The total cost of accidents where car users are involved is 5120 million €. This includes the cost where car users are victims (3715 million €) and where car users are involved (2975 million €) less the cost for car–car collision accidents (1570 million €), which otherwise would have been included twice. Dividing this total cost

Table 2. Total Accident Cost Matrix, Sweden 1997–1999 (million €).

Involved Element/Victim	All	Single Accidents	Two Party Accidents			
			Pedestrian/cyclist	*Car*	*Heavy vehicles*	*Motorcycle*
All	6,265	2,253	180	**2,975**	821	35
Pedestrian/cyclist	1,871	293	156	1,173	228	20
Car	**3,715**	1,604	14	**1,570**	517	10
Heavy vehicles	292	163	2	72	56	0
Motorcycle	387	193	8	160	20	5

Source: Lindberg (2001). Includes a b-value. Adjusted for underreporting.

Table 3. The External Marginal Cost and its Components, Sweden 1997–1999.

$MC = r(a+b+c)(1-\theta+E)+\theta(rc)$	$r(a+b+c)$	θ	E	rc
0.02 €/vkm	0.088 €/vkm	0.73	−0.10	0.01 €/vkm

with the total number of car vehicle kilometres will give information on two parts in the marginal cost expression (8). The cost share that falls on car users follows also from this general table (Table 3) and is $\theta = 0.73$ (3715/5120). The elasticity has to be found from other studies, let us assume $E = -0.10$ here. Information on the average system external cost (rc) is also shown.

4. TWO CASE STUDIES

While the result presented above is based on general information and thus not very precise in its policy prescription, this section presents two applications of the theory on a more detailed level. The first case (Section 4.1) derives an external marginal accident cost for railway level crossing accidents and focus on the relationship between traffic volume and accidents, i.e. the risk elasticity. The other case (Section 4.2) is an application on HGVs where both the problem of the proportion internal cost and the relationship between the distance driven and accidents becomes apparent.

4.1. Railway Level Crossing Accidents

Since the separation of the Swedish railway monopoly in 1988, into a Track Authority (*Banverket*) and a Railway company (*SJ*), the operator(s) have been charged for the use of the infrastructure. The charge should in principle be based on the marginal costs of infrastructure damage, including damage on overhead lines, marginal accident and environmental costs. The accident charge is today differentiated around an average of 0.90 Swedish Krona (SEK)/train.km with 1.10 SEK/train.km for passenger trains and 0.55 SEK/train.km for freight trains, which is one-third of the charge related to train kilometers.[8] The current charge does not include the cost of level crossing accidents.

Based on 5 years' data on total number of accidents (ACCID), personal injury accidents (PACCID), number of fatalities (FATAL), severe (SEVERE) and slight injuries (SLIGHT), a database on level crossing accidents has been constructed (Table 4). The location of the crossings are recorded,[9] type of protection device, type and name of the passing road and if the track is a single or double track. For each segment, the annual number of passenger (QP) and freight trains (QF) for years 1996 and 1998 have been collected from the Swedish Railways. Road traffic volume has been included manually for main roads and county roads based on information from Swedish National Road Administration. However, it was only possible to add road traffic flow for 981 of the 8600 crossings. The official Swedish accident values are used, which are comparable with the values presented in Section 2.

Per passing train, the cost varies from 0.59 SEK/train on the unprotected crossings to 7.4 SEK/train on the open crossings with St Andrew cross. The average cost is 2.3 SEK/train. The study only considers road/rail level crossing accidents and costs related to personal injuries. This means that virtually all costs fall on the road user and, consequently, we do not need to split the cost between different user groups. The key function to understand becomes the relationship between train traffic volume and accident cost (Fig. 1).

Based on literature survey and discussions with rail safety staff, a basic model structure has been developed.[10] The number of accidents will be a function of exposure and user misbehaviour. As the exposure increases, the number of accidents will increase and as the probability of misbehaviour increases, the number of accidents will increase.

The exposure can be seen as the probability that a car and a train will meet exactly on the track. This probability depends on the number of passing trains and cars as well as the speed when passing the crossing. It is also possible that the number of tracks, measuring the width of the crossing, will increase the number of accidents. Misbehaviour, mainly of road users, is much more difficult to observe. The alignment of the road before the

Table 4. Annual Accidents by Severity and Total Accident Cost per Crossing and Cost per Train Passages, 1995–1999.

Annual	P0 No Protection Device	P1 Full Barriers	P2 Half Barriers	P3 Open Crossing w. Flash Light	P4 Open Crossing w. St Andrew Cross[a]	All
Accidents per crossing						
FATAL	0.00013	0.00051	0.00199	0.00144	0.00268	0.00084
SEVERE	0.00009	0.00051	0.00020	0.00058	0.000535	0.00026
SLIGHT	0.00018	0.00068	0.00279	0.00260	0.001784	0.00095
PACCID	0.00031	0.00135	0.00299	0.00375	0.003568	0.00146
ACCID	0.00062	0.05616	0.04726	0.01241	0.0067797	0.01546
Accidents per million passing train						
PACCID/ (QP + QF)	0.085	0.137	0.341	0.842	0.665	0.271
ACCID/ (QP + QF)	0.170	5.710	5.392	2.785	1.263	2.856
Accident cost per passing train (SEK/train)						
AC	0.586	0.875	3.337	5.023	7.389	2.344

[a]Or minor private road light.

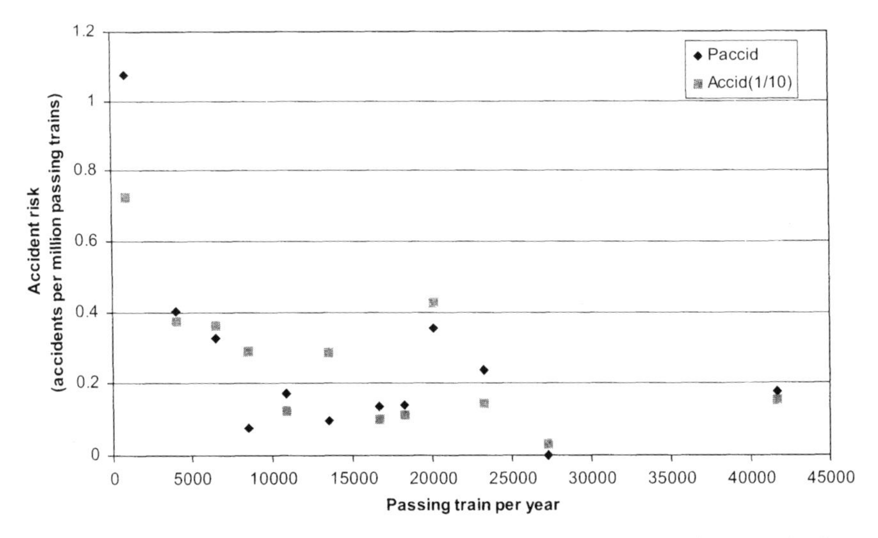

Fig. 1. Interface Accident Risk, all Accidents (divided with 10) and Personal Injury Accidents only 1995–1999.[13]

crossing has probably a correlation to this but we do not have any information of this. Train speed increases the risk of misbehaviour. Given the same speed, we do not expect to find different risks for passenger and freight trains.

Two main groups of accident models are estimated. In Group A only crossings with observed road traffic volume have been included while Group B covers all crossings. As a proxy for road traffic volume road type is used.

Accidents are unlucky events that depend on a number of known and unknown reasons. For one individual crossing the annual risk that an accident will happen is a certain probability. The probability (P) that an accident will happen can be written as $P = A/Y$ where A is the number of (personal injury) accidents and Y the number of crossings. With a logit model this probability can be expressed as (9), where x is the independent variable, β the parameters that will be estimated, and $\Lambda(\beta' x)$ indicates the logistic cumulative distribution function.

$$P(A = 1) = \frac{e^{\beta' x}}{1 + e^{\beta' x}} = \Lambda(\beta' x) \tag{9}$$

Both the volume of trains (QTOT98 = QP + QF, year 1998) and the logarithm of road vehicles (LNQR) explain significantly the number of personal injury accidents (PACCID) with expected sign; more trains or cars will increase the number of expected accidents. However, in the initial model A0, as seen in Table 5 neither the proxy for train speed (SPEED),[11] car speed (R1R2-main and county roads) nor the crossing width (DOUBBLE = 1) are significant. The existence of urban roads (R3) increases the risk, while barriers (P1) and half barriers (P2) reduce the risk as expected. After reducing insignificant variables in a number of steps, a second model (A1) presents significant effects for urban road (R3), full (P1) and half barriers (P2). A Wald test shows that the model cannot distinguish between half and full barriers. The final model in Group A (A2) only includes train and road traffic and the existence of barriers (full or half, P1P2).

In Group B, Table 6, road type is used as a proxy for road traffic volume. Train traffic volume (QTOT98) affects the accident probability, and it appears that road type works as a proxy for road traffic volume. The existence of barriers (P1P2) reduces the risk. A Wald test suggests that the model cannot distinguish between the detailed road categories in model B0. An aggregate of main and county roads (R1R2) and streets and other roads (R3R4) suggests that the former roads have a higher risk than the latter smaller roads (model B1). The Group B models predict the observed probability less accurate than Group A.

The marginal effect on the probability can be written as (10) (see Greene, 1993). The marginal effects can also be expressed in terms of risk elasticity E as follows from (11) and (12) below.

$$\frac{\mathrm{d}P}{\mathrm{d}x} = \Lambda(\beta' x)(1 - \Lambda(\beta' x)\beta) \tag{10}$$

$$r = \frac{A}{Q} = \frac{P}{Q} Y \tag{11}$$

$$E = \frac{\partial R}{\partial Q}\frac{Q}{r} = \frac{\mathrm{d}P}{\mathrm{d}Q}\frac{Q}{P} - 1 \tag{12}$$

The annual (as compared to the 5-year effect) marginal effect of train volume on the probability ($1/5\,\mathrm{d}P/\mathrm{d}Q$) is 1.24×10^{-7} for model A2, which includes road traffic volume. For B1, with a proxy for road traffic volume, the marginal effect is 0.40×10^{-7}. We evaluate the marginal effect for all crossings (Pall) and for different protection devices. The presented elasticities are based on the predicted annual probability (P/5 model). The model

Table 5. Estimated Accident Models, Group A.

Variable	MODEL A0		MODEL A1		MODEL A2	
	Parameter	t ratio	Parameter	t ratio	Parameter	t ratio
Constant	−7.097	−4.420**	−6.026	−4.809**	−5.844	−4.692**
LNQR	0.536	2.712**	0.471	2.4764**	0.447	2.360**
QTOT98	0.543×10^{-4}	1.680*	0.490×10^{-4}	2.316**	0.460×10^{-4}	2.231**
SPEED	−0.299	−0.424				
DOUBBLE	0.209	0.165				
R1R2	1.089	1.304				
R3	3.165	2.234**	1.859	1.590		
P1	−2.901	−3.676**	−2.571	−3.544**		
P2	−1.783	−2.758**	−1.473	−2.580**		
P1P2					−1.885	−3.482**
Wald Statistics	BR1R2−βR3 = 04.456 BP1−βP2 = 0		βP1−βP2 = 0	2.596		
Log-L	−82.55		−84.31		−86.33	
Log-L(0)	−93.57		−93.7		−93.8	
LR	22.05		18.72		14.84	
LRI	0.118		0.100		0.079	
dP/dQ	6.065×10^{-7}	1.641	5.98×10^{-7}	2.212**	6.19×10^{-7}	2.22**
$1/5\,dP/dQ$	1.21×10^{-7}		1.20×10^{-7}		1.24×10^{-7}	

*Significant at 10%.
**Significant at 5%.

Table 6. Estimated Models, Group B.

Variable	MODEL B0		MODEL B1	
	Parameter	*t* ratio	Parameter	*t* ratio
Constant	−6.517	−14.396**	−6.500	−14.369**
QTOT98	3.97×10^{-5}	3.248**	3.65×10^{-5}	3.008**
R1	3.945	4.329**		
R2	2.772	4.814**		
R3	1.411	2.193**		
R4	2.101	4.323**		
P1P2	−0.863	−2.442**	−0.864	−2.514**
R1R2			2.871	5.062**
R3R4			1.983	4.112**
Wald statistics	βRx−βRy = 0	Not significant	βR1R2−βR3R4 = 0	6.706**
Log-L	−327.821		−329.869	
Log-L(0)	−351.914		−351.914	
LR	48.19		44.09	
LRI	0.068		0.063	
d*P*/d*Q*	2.12×10^{-7}	2.947**	2.00×10^{-7}	2.762**
d*P*/d*Q* 1/5	0.423×10^{-7}		0.400×10^{-7}	

**Significant at 5%.

A2 is based on a subset of the data where crossings with full (P1) and half barriers (P2) dominate. The marginal effect is only significant for these two protection types; the elasticity E_A is −0.65 for P1, −0.71 for P2 and −0.68 for an aggregate of barriers (P1P2) as seen in Table 7. For the B model (B1), the marginal effect is of the same magnitude; the elasticity E_B is −0.72 for the aggregate of barriers (P1P2). For other barriers the elasticity is lower; for the aggregate of open crossings with light or St Andrew cross $E_B = -0.85$ and for unprotected crossings $E_B = -0.92$. The overall elasticity is −0.75 for model A and −0.87 for model B.

We use three approaches to estimate the marginal cost. First, the *estimated marginal effect* from model B is used (see Table 8). In a second approach, the *observed probability* is linked with the estimated *elasticity from model B*. The model B overestimates the average probability for Pall, and for P0 (and P1), while it is underestimated for P3P4 (and P2, P3, P4). These differences are mirrored in the marginal costs. The third approach uses the *estimated elasticity* from model A (E_A) with the *observed probabilities* for the whole database.

Table 7. Accident Probability and Marginal Effect.

Model A2	Pall	P1	P2	P1P2	P3	P4	P3P4	P0
1/5 dP/dQ (10^3)	0.00012**	0.00011**	0.00009**	0.00010**	0.00033*	0.00039*	0.00034*	0.00020[a]
P/5 model (10^3)	3.88	3.10	2.36	2.74	8.64	10.90	8.98	5.38
P/5 observed (10^3)	3.88	1.96	3.56	2.74	9.28	7.70	9.04	0.00
QTOT98	7789	9717	8088	8919	2428	4380	2715	3579
E_A	*−0.75*	*−0.65*	*−0.71*	*−0.68*	(*−0.91*)	(*−0.84*)	(*−0.90*)	(*−0.87*)
Model B1	**Pall**	**P1**	**P2**	**P1P2**	**P3**	**P4**	**P3P4**	**P0**
1/5 dP/dQ (10^3)	0.00004**	0.00007**	0.00007**	0.00007**	0.00011**	0.00007**	0.00008**	0.00002**
P/5 model (10^3)	1.69	2.28	2.13	2.21	2.04	2.28	2.66	0.93
P/5 observed (10^3)	1.46	1.34	3.00	2.10	3.76	3.56	3.64	0.30
QTOT98	5415	9834	8765	9344	4455	5367	5019	3654
E_B	*−0.87*	*−0.71*	*−0.72*	*−0.72*	*−0.76*	*−0.84*	*−0.85*	*−0.92*

[a]Significant at 15%.
*Significant at 10%.
**Significant at 5%.

Table 8. Marginal Cost by Passing Train (SEK/Passage).

	All	P1	P2	P1P2	P3	P4	P3P4	P0
MC model B	0.35	0.58	0.58	0.58	0.94	0.58	0.70	0.18
MC r_obs E_B	0.30	0.34	0.82	0.55	1.73	0.91	0.96	0.06
MC r_obs E_A	(0.58)	0.42	0.87	0.62	(0.69)	(0.89)	(0.65)	(0.10)

The marginal costs based on observed probabilities and model B elasticities are 0.30 SEK/passage for all crossings, 0.55 SEK/passage for barriers (P1P2), 0.96 SEK/passage for open crossings with light or St Andrew cross (P3P4) and 0.06 SEK/passage for unprotected crossings. With model A elasticity the marginal cost at barriers are 13% higher. We cannot distinguish the marginal cost for freight and passenger trains separately, although we expect that slower freight trains have a lower marginal cost.

The results provide a strong case for charging rail operators for the accident costs imposed on road users at rail level crossings. In terms of economic efficiency, this argument holds regardless of whether the accidents are due to the negligence of road users.

If the charge is differentiated according to protection device, new incentives will be created in the railway safety system. If the number of dangerous crossings can be reduced, the charge will go down.

4.2. Heavy Goods Vehicles

Marginal cost of HGVs has previously been dominated by the cost of road wear and tear. This cost increases exponentially with axle weight, the so-called "forth-power law". Consequently, road taxes and charges based on marginal cost theory increase strongly with axle weight. As the accident cost component becomes more important, the structure of the accident component becomes crucial; does the external accident cost increase with axle weight and reinforce the current structure on taxes or charges or does it decrease with axle weight, i.e. cancelling out the axle dependence of the tax structure.

This case study has two basic data sources, information on individual accidents during 1999 from the Swedish National Road Administration (*Vägverket*) and information on the distance driven during 1999 by goods vehicles above 3.5 tonne from the Swedish Vehicle Inspection Authority (*Bilprovningen*). The two sources are linked together and a database with distance driven for all 78,000 goods vehicles and their individual accident record are constructed. The database includes 83 fatalities, 254 severe injuries, and 1,035 slight injuries. This is 5.8% of all the (reported) accidents in Sweden during 1999. In general, goods vehicle accidents are more severe than the average road accident (14% of the fatalities) even if they include less passengers and unprotected users (bicyclist, moped users (1.2%) or pedestrians (2.6%)). The production year of the vehicle (YEAR), the total weight (WEIGHT), number of axles (AXLE), body type and result at the annual inspection have been included in the database.

The study focuses on distance driven by individual vehicles, which is a problem. This is different from the traffic volume on a single road. We may expect that vehicles driven longer distances drive more on interurban roads, have a more experienced driver and may be better maintained. While we may expect that a higher share of HGVs on a single road increase the risk, we may expect longer distances driven to decrease the risk. Our analysis may understate the risk resulting from an equi-proportionate increase in traffic on all roads.

The characteristics of HGV accidents are different from passenger car accidents mainly because of the special attributes of these vehicles; they are much heavier and larger in dimensions than passenger cars; they have less effective acceleration than passenger cars and have problems to maintain their speed on uphill, and they have lower deceleration in response to braking than have passenger cars. The result is that HGV accidents often are more serious than other road accidents, especially for the non-HGV road users. Improved technology should improve the safety performance and the risk is expected to increase with vehicle age.

The reason *why* HGV accidents should differ from passenger car accidents also means that we should expect a difference in accident rates for different weight classes of HGVs. However, a survey of the literature did not result in any conclusive evidence on the relationship between truck configuration and accident risk (see e.g. Nilsson (1996)). National Highway Traffic Safety Administration (1992) concludes that single-unit trucks give rise to more injuries in collisions than tractor–trailer combinations. However, this could possibly be explained by the fact that combination truck operates predominantly long-haul, high-speed, in rural environment, while single-unit trucks are used more often in an urban environment, at lower speeds, in daytime.

As the number of trucks increases with a *given flow of other vehicles*, we may expect different reactions on the accident probability with different vehicle elements. Jovanis and Chang (1986) report an effect on overall accidents from truck traffic in a study of the Indiana Toll Road. The total number of accidents increases at a decreasing rate as the truck traffic increases. Later studies (Joshua & Garber, 1990; Miaou, 1994) on more general data have supported the conclusion that truck traffic has an influence on the total number of accidents.

The average accident risk per goods vehicle is 41 police reported accidents per 1,000 vehicles as presented in Table 9. The vehicles are grouped into 7 weight classes, where classes 2–7 is defined as HGV, i.e. above 12 tonne. This risk is (almost) strictly increasing with weight class where the lightest vehicles (class 1) have 1/3 of the average risk and the heaviest vehicles (class 7) have a risk 57% above the average risk. The highest risk per vehicle, more than twice as high as the average risk, can be found in weight class 6 (27–31 tonne). The accident risk per vehicle kilometre does not show the same clear pattern. The average risk is 0.83 police reported accidents per million-vehicle kilometre. The lightest vehicles (class 1) have a risk per kilometre that is 75% of the average risk, and the heaviest vehicles (class 7) have a risk 23% above the average risk. The less clear relationship between

Table 9. Accident Risk per Registered Vehicle and per Kilometre by Weight Class, Sweden, 1999.

Weight Class	Weight Class	Min. Weight (kg)	Max. Weight (kg)	Accidents During 1999	Accidents per Vehicle	Accidents per Million Vehicle Kilometre
1	1	3,510	11,990	204	0.0139	0.633
2	2	12,000	14,990	70	0.0229	1.002
3	3	15,000	18,990	225	0.0293	0.896
4	4	19,000	22,990	212	0.0482	0.724
5	5	23,000	26,980	754	0.0592	0.977
6	6	27,000	30,970	517	0.0837	0.914
7	7	31,000	159,460	74	0.0647	1.030
All	All			2055	0.0412	0.837
All>12				1854	0.0527	0.869

risk and weight when we study risk per vehicle kilometre could be due to differences in the type of exposure as suggested by the National Highway Traffic Safety Administration (1992); heavier vehicles operate predominantly on rural roads.

Only 23% of the killed and injured persons in accidents involving a goods vehicle were drivers or passengers of the goods vehicle. The majority of the victims were drivers of other motor vehicles (57%) or passengers in other motor vehicles (18%). Only 3% were unprotected road users during 1999. For fatalities, the imbalance between the number of victims in the goods vehicle and in other vehicles is even more striking; only 6% of the killed persons belonged to the HGV.

The total cost of personal injuries in police-reported accidents involving goods vehicles were 224 million € in 1999. Only 9% of the accident cost falls on the goods vehicle users, 77% on other motor vehicle users and 2% on unprotected users; 11% is so-called system external costs and falls on the society in general; it consists mainly of the hospital and medical costs paid by the general social security system.

The average cost of a goods vehicle accident was 57,000 € with 13,000 € per accident for weight class 1 and 94,000 € per accident for weight class 6. The proportion of internal cost has been calculated in relation to the total cost, excluding the system external cost, ($\theta = \text{IC}/(\text{IC} + \text{EVC} + \text{EUC})$). The proportion of internal cost is between 0.03 and 0.38 for the different weight classes. The proportion of internal cost does not show any clear relationship with weight class, although the highest proportions can be found for the lightest vehicles and the lowest proportion for the heaviest vehicles (Table 10).

A similar binary choice model as for railway accidents has been estimated. In Table 11, the result is presented for three models. The first two models are based on the whole sample of goods vehicles while the last model is based on HGV, i.e. only vehicles above 12 tonne. The weight, tonnes per axle and driven distance increase the probability for an accident as expected. A remark in the annual inspection increases also the accident probability and we find a higher probability for delivery vehicles (HGV2). However, model 1 suggests that the accident probability declines with increased age. Therefore, age is omitted from model 2.

We are mainly interested in the marginal effects with respect to driven distance, which are presented in Table 12 for models 2 and 3.

The estimated marginal effect and calculated elasticity suggest that the number of accidents increases with the number of kilometres driven. However, the number of accidents does not increase in proportion to the increase

Table 10. Internal and External Costs per Goods Vehicle Accident in Sweden, 1999 by Weight Class (kEuro/Accident).

Weight Class	Internal Cost IC (kEuro)	External Cost –other vehicles EVC (kEuro)	External Cost – unprot. user EUC (kEuro)	System External Cost ECC (kEuro)	Total External Cost TEC (kEuro)	Total Cost TC (kEuro)	PROP(θ)
1	11.5	26.9	0.0	4.9	31.9	43.4	0.30
2	4.6	24.5	1.0	6.1	31.6	36.2	0.15
3	4.7	63.3	0.9	8.1	72.3	77.0	0.07
4	3.5	37.0	0.0	5.5	42.5	45.9	0.09
5	5.9	42.6	0.4	6.1	49.1	55.0	0.12
6	3.4	43.8	4.5	6.0	54.2	57.6	0.07
7	2.8	86.0	0.0	10.5	96.5	99.3	0.03
All	5.3	43.9	1.4	6.3	51.5	56.8	0.10
HGV	4.6	45.7	1.5	6.4	53.7	58.3	0.09

Table 11. Estimated Coefficients for all Goods Vehicles.

	MODEL 1		MODEL 2		MODEL 3	
	Coefficient	t value	Coefficient	t value	Coefficient	t value
Constant	−4.457	−30.268	−5.689	−43.285	−6.05834	−21.283
AGE	−0.0698	−14.131	—	—	—	—
WEIGHT	0.0000403	7.748	0.0000531	10.344	0.0000547	10.357
TPA	0.0000761	3.75	0.000115	5.568	0.00015033	4.609
DIST99	0.0000037	14.209	0.00000469	19.223	0.00000485	17.558
INSPOK	0.292	5.839	0.159	3.234	0.166	3.186
HGV2	0.216	4.243	0.347	6.945	0.294	5.489
Log-L	−7871.03	—	−7880.72	—	−6935.80	—
Log-L(0)	−8571.84	—	−8571.84	—	−7254.57	—
LR	1401.6	—	1382.2	—	637.5	—
LRI	0.082	—	0.081	—	0.044	—

Table 12. Marginal Effects with Respect to Distance (dP/dQ).

Weight Class	MODEL 2		MODEL 3	
	Coefficient	t value	Coefficient	t value
1	0.053×10^{-6}	12.135	—	—
2	0.097×10^{-6}	17.547	0.090×10^{-6}	11.10
3	0.142×10^{-6}	17.652	0.141×10^{-6}	15.77
4	0.210×10^{-6}	15.463	0.220×10^{-6}	14.64
5	0.245×10^{-6}	18.754	0.249×10^{-6}	16.95
6	0.315×10^{-6}	17.569	0.330×10^{-6}	16.40
7	0.344×10^{-6}	15.632	0.362×10^{-6}	15.00
All > 12	—	—	0.213×10^{-6}	17.43

in distance; as the distance increases by 10%, the number of accidents will increase by only 2%. This means that the accident risk, i.e. number of accidents per kilometre, will decline.[12] Based on this and the external and internal costs per accident presented above, the external marginal accident cost can be estimated (see Table 13).

However, the model estimates a declining risk while it has been assumed that the user internalises the average risk (PMC). This means that the user "overreacts" in relation to his or her own changed risk and the model corrects for this reaction. This may be a questionable result. In Table 13, we

Table 13. External Marginal Cost by Weight Class (€/1,000 vkm).

Weight Class	MODEL 2			MODEL 3		
	MCI	*MCII*	***MCtot***	*MCI*	*MCII*	***MCtot***
1	1.69	−6.14	−4.45	—	—	—
2	3.07	−4.11	−1.03	2.84	−3.65	−0.81
3	10.27	−4.14	6.13	10.20	−3.96	6.24
4	8.93	−1.89	7.04	9.36	−1.91	7.45
5	12.03	−4.29	7.74	12.24	−4.17	8.07
6	17.08	−1.77	15.31	17.89	−1.76	16.13
7	33.23	−2.83	30.40	34.92	−2.89	32.03
HGV	—	—	—	11.41	−3.02	8.39

have decomposed the external marginal accident cost into two components; the first (MCI) is the cost imposed on other users and the second the changed cost for the own user group (MCII). The first part, i.e. the cost imposed on other users, is always positive, as longer distances are travelled, the risk for other users increases, although at a decreasing rate. However, the second part is negative. As the distance increases, the risk is reduced also for the goods vehicle user while it has been assumed that he or she internalises the average internal cost. Consequently, an externality arises, which is negative. This reduction gets more dominant when the proportion of internal cost increases, which it does for lighter vehicles.

The second part is based on a rather uncertain assumption. An alternative assumption would be that the users internalise the marginal effect, and not the average effect, of their own risk. The external marginal accident cost consists then only of MCI in the table above. The external marginal accident cost for HGV, based on driven distance, would be 11.4 €/1000 vkm with the highest cost for the heaviest goods vehicles (35 €/1000 vkm) and the lowest cost for the lightest goods vehicles.

5. SOME FURTHER KEY ISSUES

5.1. Risk-Avoiding Behaviour

Most of our empirical works suggest indeed that the risk decreases with traffic volume ($E<0$). The reduction in the accident risk is so large in some cases that it offsets the external cost related to other categories. This highlights one of the problems of the presented approach – risk-avoiding behaviour.

The user may react in a number of different ways when he perceives that the risk level has changed. Peltzman (1975) developed the hypothesis of *risk compensation* and presented evidence showing that the user when given a safer environment compensates this with a higher degree of risk taking. In the same way, a more unsafe environment may be compensated by the user with reduced exposure to risk. This reaction generates a cost to the user and reduces the observed change in risk. The cost of this *risk-avoiding behaviour* has to be included in the external marginal cost.

In the following we divide the users into two groups, the first group (A) is injured and the second (B) is the other part in the accident. We assume that the level of safety (s) on a given trip is associated with a cost (g), which increases as the level of safety increases. The total annual cost (TC) for accidents and traffic safety can then be written as

$$\mathrm{TC} = A(a + b + c) + Q_{\mathrm{A}}g(s_{\mathrm{A}}) + Q_{\mathrm{B}}g(s_{\mathrm{B}}) \tag{13}$$

We allow only for risk-avoiding behaviour that increases the "internal" safety, i.e. the user's own safety. Users who expect to be victims (user A; e.g. unprotected road users) will adjust behaviour (s_{A}) to protect themselves from an accident. To the marginal accident cost, the cost of all victims' risk-avoiding behaviour has to be added. The total external cost for category B, the unharmed user ($\theta = 0$) will be

$$MC_{\mathrm{B}}^{e} = r_{\mathrm{B}}(a + b + c)[1 + E_{\mathrm{B}}] + Q_{\mathrm{A}}\frac{\mathrm{d}g}{\mathrm{d}s_{\mathrm{A}}}\frac{\mathrm{d}s_{\mathrm{A}}}{\mathrm{d}Q_{\mathrm{B}}} \tag{14}$$

The last term in Eq. (14) is the cost of user group A's risk-avoiding behaviour triggered by an increased number of trips of category B. A part of the risk-avoiding behaviour, lower speed, can be traced to the congestion cost and handled as such. Another part can be found in infrastructure cost, where a higher number of, for example, flights increases the necessary number of safety staff and the dimension of the rescue capacity at the airport.

Peirson, Skinner, and Vickerman (1994) introduces a form of risk-avoiding behaviour where the users, when they selfishly adapt their behaviour, reduce the risk for all other users. The risk-avoiding behaviour includes an element of positive externality. Johansson (1996) shows how this can be internalised through the accident externality charge in a second-best situation where the behaviour is not subsided per se.

5.2. Liability

The theory presented in Section 3 does not explicitly discuss liability. As above, we assume that the first group (A) is injured and the second (B) is the other party in the accident. Without any liability, the injured user (A) will bear all costs and the other part (B) will not bear any cost. The external marginal cost for each group is then:

$$\mathrm{MC}_{\mathrm{A}}^{\mathrm{e}} = r_{\mathrm{A}}(a+b+c)[1+e_{\mathrm{A}}] - r_{\mathrm{A}}(a+b) = r_{\mathrm{A}}(a+b+c)[E_{\mathrm{A}}] + r_{\mathrm{A}}c \quad (15)$$

$$\mathrm{MC}_{\mathrm{B}}^{e} = r_{\mathrm{B}}(a+b+c)[1+E_{\mathrm{B}}] \quad (16)$$

The final expression of Section 3 (Eq. (8)) is a weighted sum of these two expressions, where θ expresses the probability of being the injured user.

Under a negligence rule, users in Group B, will not bear any cost as long as they behave legally. If they break the law, they will be responsible for some of the costs as compensation (d) to user A or as a fine (M). These costs will be included in their private marginal cost and the external cost will decrease. At the same time, the compensation (d) will reduce the expected cost of an injured user in Group A; consequently the external marginal cost of Group A will increase. While this conclusion at first looks disturbing, it should be noted that the criminal user B will have a higher *generalised cost,* than the legal user B. The criminal user has to pay fine, compensation and external marginal cost.

$$\textit{Legal user}\ \mathrm{B} \Rightarrow MC_{\mathrm{B}}^{\mathrm{e}} = r_{\mathrm{B}}(a+b+c)[1+E_{\mathrm{B}}] \quad (17)$$

$$\textit{Legal user}\ \mathrm{B} \Rightarrow MC_{\mathrm{B}}^{\mathrm{e}} = r_{\mathrm{B}}(a+b+c)[1+E_{\mathrm{B}}] - r_{\mathrm{B}}(d+M) \quad (18)$$

$$\textit{Not compensated user A} \Rightarrow MC_{\mathrm{A}}^{\mathrm{e}} = r_{\mathrm{A}}(a+b+c)[E_{\mathrm{A}}] + r_{\mathrm{A}}c \quad (19)$$

$$\textit{Compensated user}\ \mathrm{A} \Rightarrow MC_{\mathrm{A}}^{\mathrm{e}} = r_{\mathrm{A}}(a+b+c)[E_{\mathrm{A}}] + r_{\mathrm{A}}(c+d) \quad (20)$$

With strict liability for user B, he or she will always pay the cost in the form of compensation (d) or a fine (M) in the case of an accident – in principle he or she is always a "criminal" as in Eq. (18).

Assume that both A and B are car users and that we cannot ex ante identify the criminal user; we have to assume that the probability of being in either group is 50/50. Consequently, while the marginal cost of Group B is reduced, it is increased for Group A through the compensation (d); the effect on the joint marginal cost for all car users disappears and the external marginal cost can be written once again as in Section 3. However, a fine (M)

will affect the result. This theory assumes that users perceive the compensation and fine as a part of their cost ex ante.

An interesting case is if the victim (user A) is guilty of the accident, for example, a pedestrian who crosses the street illegally and is hit by a car. Depending on the legal situation the car driver should ex ante be charged as Eq. (17) or (18) and the pedestrian as (19). Consequently, the innocent car driver shall pay a charge ex ante. Is this an ethical problem? It may be. However, if it is too expensive to ensure that the pedestrian A behaves legally, it may be optimal to reduce the number of trips of the injurer B. While this conclusion sounds disturbing, it is less obvious with another example. If it is too expensive to ensure that children behave legally around the playground, it is more efficient to reduce the level of car traffic on the adjacent streets.

5.3. Charges on Behaviour?

The number of trips is a part of the traffic safety problem. Internalisation of the external marginal cost, that has been discussed, ensures an optimal number of trips. In addition, if the marginal cost is estimated for different road types or road/rail crossing protection devices, a differentiated charge will also influence the route choice.

However, traffic safety policy is not only, or even mainly, about optimal traffic volumes; more important is probably a safe behaviour while driving – this will not be influenced by a charge based on the external marginal accident cost measured related to driven distance. A more sophisticated system may observe the actual behaviour of the driver and can differentiate the charge appropriately.

NOTES

1. Or willingness-to-accept (WTA) an increased risk.
2. Needleman (1976); Jones-Lee (1992); Schwab Christe and Soguel (1995) and Lindberg (1999).
3. Slovic, Fischoff, and Lichtenstein (1982, p. 467).
4. See Viscusi (1998) for a short discussion on this topic.
5. Viscusi Hakes and Carlin (1997).
6. Benjamin, Dougan, and Buschena (2001).
7. Our assumption is close to the risk homeostasis theory (Wilde, 1981) where the user acts as a utility maximiser and tries to keep his risk in equilibrium with his target risk.

8. The charge for infrastructure damage is approximately 3.50 SEK/train.km for an average train of 495 tonne/train.

9. By track-segment, distance in kilometre and metre from the beginning of the segment.

10. Coleman and Stewart (1977), Coulombre, Poage, and Edwin (1982), Hauer and Persaud (1987), Highway Research Board (1968) and TFD (1981, 1983).

11. Proportion passenger train.

12. The risk elasticity is −0.83 for model 2. This is in line with the elasticity reported by Jovanis and Chang (1986) ($E = -0.8$).

13. Crossings are grouped by number of passing trains. The first eleven groups have an interval of 2,500 passing train. The number of observations in each group are 3,382; 2,229; 941; 652; 419; 459; 264; 77; 28; 36; 23. The 12th and last group contains crossings with 30,000–75,000 passing trains (108 obs).

REFERENCES

Ardekani, S., Hauer, E., & Jarnei, B. (1997). Traffic impact models. In: Gartner, N., Messers, C. & Rathi, A. (Eds.), *Traffic flow theory – A state-of-the-art report*. Oak Ridge National Laboratory, Turner-Fairbank Highway Research Center. http://www.tfhrc.gov/its/tft/tft.htm

Beattie, J., Covey, J., Dolan, P., Hopkins, L., Jones-Lee, M., Loomes, G., Pidgeon, N., Robinson, A., & Spencer, A. (1998). On the contingent valuation of safety and the safety of contingent valuation: Part 1-caveat investigator. *Journal of Risk and Uncertainty, 17*, 5–25.

Benjamin, D. K., Dougan, W. R., & Buschena, D. (2001). Individuals' estimates of the risk of death: Part II – new evidence. *Journal of Risk and Uncertainty, 22*(1), 35–57.

Bergstrom, C. T. (1982). When is a man's life worth more than his human capital. In: Jones-Lee (Ed.), *The value of life and safety*. Amsterdam: North-Holland Publishing Company.

Blaeij, A. (2003). The value of statistical life: A meta analysis. *Accident Analysis and Prevention, 35*, 973–986.

Carthy, T., Chilton, S., Covey, J., Hopkins, L., Jones-Lee, M., Loomes, G., Pidgeon, N., & Spencer, A. (1999). On the contingent valuation of safety and the safety of contingent valuation: Part – 2 the CV/SG 'chained' approach. *Journal of Risk and Uncertainty, 17*(3), 187–213.

Chambron, N. (2000). Comparing six DRAG-type models. In: M. Gaudry & S. Lassarre (Eds), *Structural road accident models – The international drag family* (pp. 205–224). Amsterdam: Elsevier.

Coleman, J., & Stewart, G. (1977). Investigation of accident data for railroad–highway grade crossings. *Transportation Research Record, 611*, 60–67.

Coulombre, R., Poage, J., & Edwin, H. F. (1982). *Summary of the DOT rail-highway crossing prediction formulas and resource allocation model*. U.S. Department of Transportation Federal Research and Special Programs Administration, Cambridge, September 1982.

Dickerson, A. P., Peirson, J. D., & Vickerman, R. W. (2000). Road accidents and traffic flows: An econometric investigation. *Economica, 67*, 101–121.

ECMT. (1998). *Efficient transport for Europe, policies for internalisation of external costs*. Paris: European Conference of Ministers of Transport.

Fridstrøm, L., Ifver, J., Ingebrigtsen, S., Kulmala, R., & Thomsen, L. K. (1995). Measuring the contribution of randomness, exposure, weather, and daylight to the variation in road accident counts. *Accident Analysis and Prevention, 27*, 1–20.

Greene, W. H. (1993). *Econometric analysis* (2nd ed.). New York: Macmillan Publishing Company.

Hauer, E., & Persaud, B. N. (1987). How to estimate the safety of rail-highway grade crossings and the safety effects of warning devices. *Transportation Research Record, 1114*, 131–140.

Hauer, E., & Bamfo, J. (1997). In: Proceedings of the ICTCT Conference, *Two tools for finding what* function *links the dependent variable to the explanatory variables*, Lund.

Highway Research Board (1968). *Factors influencing safety at highway-rail grade crossings*. National Cooperative Highway Research Program Report No. 50.

Hochman, H. D., & Rodgers, J. D. (1969). Pareto optimal redistribution. *American Economic Review, 59*(4), 542–557.

Jansson, J. O. (1994). Accident externality charges. *Journal of Transport Economics and Policy, 28*(1), 31–43.

Johansson, O. (1996). *Welfare, externalities, and taxation; theory and some road transport applications*. Göteborg: Gothenburg University.

Jones-Lee, M. W. (1990). The value of transport safety. *Oxford Review of Economic Policy, 6*(2), 39–58.

Jones-Lee, M. W. (1992). Paternalistic altruism and the value of statistical life. *Economic Journal, 102*, 80–90.

Joshua, S. C., & Garber, N. J. (1990). Estimating truck accident rates and involvement using linear Poisson regression models. *Transportation Planning and Technology, 15*, 41–58.

Jovanis, P. P., & Chang, H. L. (1986). Modelling the relationship of accidents to miles travelled. *TRR, 1068*, 42–58.

Kahneman, D., & Knetch, J. (1992). Valuing public goods. The purchase of morals satisfaction. *Journal of Environmental Economics and Management,, 22*, 57–70.

Lindberg, G. (1999). *Benevolence and the value of statistical life.* Working Paper CTEK: Borlänge.

Lindberg, G. (2001). Traffic insurance and accident externality charges. *Journal of Transport Economics and Policy, 35*(3), 399–416.

Miaou, S. P. (1994). The relationship between truck accidents and geometric design of road sections: Poisson versus negative binomial regressions. *Accident Analysis and Prevention, 25*(6), 689–709.

National Highway Traffic Safety Administration. (1992). *Heavy-duty trucks in crashes NASS 1979–1986*. Washington: US Department of Transportation.

Needleman, L. (1976). Valuing other people's lives. *Manchester School, 44*, 309–342.

Newbery, D. M. (1988). Road user charges in Britain. *Economic Journal, 98*, 161–176.

Nilsson, G. (1996). *Trafiksäkerhetssituationens variation i tiden.* VTI Rapport 15.

Peirson, J., Skinner, I., & Vickerman, R. (1994). *The microeconomic analysis of the external costs of road accidents*. Discussion paper 94/6, CERTE, May 1994.

Peltzman, S. (1975). The effects of automobile safety regulations. *Journal of Political Economy, 83*(4), 677–725.

Persson, U., Hjalte, K., Nilsson, K., & Norinder, A. (2000). *Värdet av att minska risken för vägtrafikskador – Beräkning av riskvärden för dödliga, genomsnittligt svåra och lindriga skador med Contingent Valuation metoden*. Lunds Tekniska Högskola, Institutionen för Teknik och samhälle, Bulletin 183.

Schwab Christe, G. N., & Soguel, C. N. (1995). *Contingent valuation, transport safety and the value of life*. Dordrecht: Kluwer Academic Publishers.

Slovic, P., Fischoff, B., & Lichtenstein, S. (1982). Facts versus fears: Understanding Perceived Risk. In: D. Daniel Kahnemann, P. Slovic & A. Tversky (Eds), *Judgement under uncertainty: Heuristics and biases*. Cambridge: Cambridge University Press.

TFD (1981). *Olyckor i plankorsningar mellan väg och järnväg*. TFD S 1981:4, Stockholm.

TFD (1983). *Olyckor i plankorsningar mellan väg och järnväg. Förarbeteende i korsningar med ljus- och ljudsignaler*. TFD S-publikation 1983:2, Stockholm.

Vickrey, W. (1968). *Automobile accidents, tort law, externalities, and insurance*. Law and Contemporary Problems, Summer.

Viscusi, W. K. (1998). *Rational risk policy*. Oxford: Clarendon Press.

Viscusi, W. K., & Aldy, J. (2003). The value of a statistical life: A critical review of market estimates throughout the world. *The Journal of Risk and Uncertainty*, *27*(1), 5–76.

Viscusi, W. K., Hakes, J. K., & Carlin, A. (1997). Measures of mortality risk. *Journal of Risk and Uncertainty*, *14*, 213–233.

Vitaliano, D. F., & Held, J. (1991). Road accident external effects: An empirical assessment. *Applied Economics*, *23*, 373–378.

Wilde, G. J. S. (1981). *Objective and subjective risk in drivers' response to road conditions. The implications of the Theory of Risk Homeostasis for accident etiology and prevention. Studies for safety in Transport*. Kingston, Ontario: Queen's University.

ENVIRONMENTAL COSTS

Peter Bickel, Stephan Schmid and Rainer Friedrich

1. INTRODUCTION

Transportation activities impose considerable environmental costs, covering a wide range of different damages, including the various impacts of emissions of noise and a large number of pollutants on human health and amenity, materials, ecosystems, flora and fauna. These environmental costs are, to a large extent, external as they are not reflected in the price paid and, hence, in the decisions made by the transport user.

Most early studies on transport externalities followed a top-down approach, giving average costs for the whole country rather than marginal costs for specific circumstances. The basis for the calculation in these studies is a whole geographical unit, a country for example. For such a unit the total cost due to an environmental burden (e.g. air pollution or noise) is calculated. This cost is then allocated based on the shares of total pollutant (or noise) emissions, by vehicle mileage, etc.

However, it is marginal external environmental costs that are important for pricing purposes. These vary considerably with the technology of a vehicle, train, ship or aircraft and site (or route) characteristics. And the sum of marginal environmental cost does not equal total environmental cost if marginal costs are not equal to average costs, so allocation of total costs based on vehicle type or site or route characteristics is not equivalent to calculating marginal external costs. Only a detailed bottom-up calculation allows a close appreciation of such site and technology dependence.

In the early 1990s the ExternE (External costs of Energy) European Research Network (see e.g. European Commission, 1999a, b; Friedrich

Measuring the Marginal Social Cost of Transport
Research in Transportation Economics, Volume 14, 185–209
Copyright © 2005 by Elsevier Ltd.
All rights of reproduction in any form reserved
ISSN: 0739-8859/doi:10.1016/S0739-8859(05)14007-4

& Bickel, 2001) started to develop and apply the Impact Pathway Approach (IPA), which is a bottom-up approach designed for quantification of site and technology dependent marginal costs. At first the IPA was applied for assessing impacts due to airborne emissions. Starting with the pollutant emission, through its diffusion and chemical conversion in the environment, impacts on the various receptors (humans, crops etc.) are quantified and, then, valued in monetary terms. In other words, information is generated on three levels:

I. the increase in burden (e.g. additional emissions and ambient concentration of SO_2 in $\mu g/m^3$) due to an additional activity (e.g. one additional trip on a specific route with a specific vehicle, train, ship, aircraft);
II. the associated impact (e.g. additional hospital admissions in cases); and
III. the monetary valuation of this impact (e.g. willingness-to-pay (WTP) to avoid the additional hospital admissions in Euro).

Since its first applications, the IPA for air pollutants has been extended to global warming and noise impacts. It has been applied in a number of research projects and policy application related studies, e.g. INFRAS/IWW (2000), European Commission (1998), AEA (1997). Other studies (e.g. WHO, 1999; McCubbin & Delucchi, 1996) as well looked at the chain of ambient pollutant concentrations due to the transport sector, human health impacts and monetary valuation. However, in contrast to ExternE, the pathway analysed did not include detailed modelling of vehicle emissions at specific locations, resulting in average, rather than marginal, cost estimates.

Damages due to climate change are one of the most important categories of fossil fuel emission related damages, but also among the most uncertain and controversial. First estimates were presented by Cline (1992), Fankhauser (1995), Nordhaus (1991) and Titus (1992). Tol (2001) estimated climate change impacts with a dynamic approach consistent with the ExternE methodology. Due to the high uncertainties involved in estimating damage costs due to climate change, many studies (see e.g. Link et al., 2001; INFRAS/IWW, 2000) have used abatement cost estimates. These costs can be interpreted as willingness-to-pay for achieving socially desirable and accepted reduction targets.

As modelling of exposure to noise is a challenging task, many older studies bypassed noise modelling and allocated damages to different vehicle categories, based on rough assumptions. Such estimates lead to average instead of marginal costs. ECMT (1998) gives a broad overview of studies carried out in different European countries.

The following sections will give an overview of the preferable methodology for quantifying marginal external environmental costs, give results from recent studies and explore the possibility of generalising marginal cost estimates. The results give focus on the cost categories that have been found to be most important:

- costs due to air pollutants,
- costs due to noise, and
- costs due to greenhouse gas emissions.

2. STATE OF THE ART IN QUANTIFYING MARGINAL ENVIRONMENTAL COSTS

2.1. Relevant Cost Categories

In the context of calculating marginal environmental costs the cost categories given in Table 1 are relevant.

The scale of impact of these cost categories is very different, both in space and time. Whereas airborne pollutants are mainly a problem at the local and regional (i.e. European) scales, the effects of greenhouse gas emissions are global in nature. Noise impacts are restricted to the very local scale of several hundred metres or a few kilometres from the emitting source. Effects

Table 1. Cost Categories Generally Relevant for Calculating Marginal Environmental Costs.

Environmental Cost Category	Description
Air pollution	Impacts of airborne emissions on human health natural environment building materials
Noise	Impacts of sound emissions on amenity human health
Global warming	Various impacts from greenhouse gas emissions
Soil and water pollution	Impacts of soil pollution by heavy metals, oil, etc. water pollution by de-icing salts, heavy metals, oil, etc.
Nuclear risks	Impacts of nuclear risks arising from electricity production

Source: UNITE, 2000.

of soil pollution are restricted to the range of several kilometres from the cause, whereas water pollution may affect areas in the range of up to several hundred kilometres. The same is true for nuclear risks, which in case of an accidental release may affect all of Europe and other continents.

As regards the time scale of the effects treated here, the nuclear risks have the longest time scale, according to the long lifetime of radionuclides, which may reach several thousand years. Due to the long atmospheric lifetime of the relevant species, several hundreds of years have to be taken into account when quantifying impacts of greenhouse gas emissions. According to the shorter lifetimes of the species involved, airborne pollutants have short-term effects occurring within hours (e.g. peak concentrations) as well as effects over several days (e.g. ozone episodes). Noise has the most limited effect, as it disappears soon after the emission source has disappeared. However, after pollution or noise has dissolved, the impacts of the exposure, e.g. hypertension, may occur several years after the exposure.

With regard to environmental costs not only the operation of a vehicle or vessel is relevant, but as well up- and downstream processes associated with the transport activity. For instance, if only the direct impacts of vehicle operation would be taken into account, practically no environmental effects of electric vehicles would be recorded. It is obvious, that this would be inappropriate as electricity production may cause considerable environmental burdens, depending on the type of electricity generation. In addition, the production, maintenance and disposal of vehicles and transport infrastructure might be relevant. Hence up- and downstream processes have to be considered when quantifying environmental costs. Table 2 shows the life cycle stages relevant for marginal cost analysis.

In an ideal approach, all life cycle stages causing short-run marginal costs should be taken into account, for all cost categories (i.e. vehicle use, fuel or electricity production and the marginal components of vehicle maintenance and infrastructure provision). This clearly would be too ambitious for most quantification exercises. In most cases it appears appropriate to focus on costs due to the use of a vehicle for all cost categories. Costs due to airborne pollutants and greenhouse gas emissions are relevant as well for fuel and electricity production.

Marginal environmental costs due to vehicle maintenance and infrastructure provision can be expected to be very small and will therefore usually not be covered. The cost category noise is to a high degree location specific. Considerable modelling effort is required, therefore, in most cases noise costs are only quantified for vehicle operation.

Table 2. Relevance of Life Cycle Stages for Quantification of Short-Run Marginal Costs.

	Ideal Approach	Pragmatic Approach
Vehicle production	–	–
Vehicle use	×	×
Vehicle maintenance	× [a]	–
Vehicle disposal	–	–
Fuel/electricity production	×	×
Infrastructure construction, maintenance, and disposal	× [a]	–

Source: UNITE, 2000.
[a]Only components depending on mileage.

2.2. Methodology

As mentioned above most early studies on transport externalities followed a top-down approach, giving average costs rather than marginal costs. Moreover, the impacts of transport activities are highly site-specific, as can most obviously be seen for noise: noise emitted in densely populated areas affects many people and thus causes much higher impacts than noise emitted in sparsely populated areas. Therefore, the starting point for the assessment of marginal damages is the micro level, i.e. the activity on a particular route. The marginal external costs of one additional vehicle are calculated for a single trip. A detailed bottom-up approach is required to be able to consider technology and site-specific parameters.

The IPA was designed to meet these requirements. Fig. 1 illustrates the procedure for quantifying impacts due to airborne pollutants: the chain of causal relationships starts from the pollutant emission through transport and chemical conversion in the atmosphere to the impacts on various receptors, such as human beings, crops, building materials or ecosystems. Welfare losses resulting from these impacts are transferred into monetary values. Based on the concepts of welfare economics, monetary valuation follows the approach of 'willingness-to-pay' for improved environmental quality. It is obvious, that not all impacts can be modelled for all pollutants in detail. For this reason the most important pollutants and damage categories ("priority impact pathways") are selected for detailed analysis.

This principle of modelling the burden (e.g. emissions), behaviour of the burden (e.g. dispersion), response of receptors (e.g. health damages) and monetary valuation can and should be applied for all impact categories. The main bottleneck of this procedure is the availability of the models required

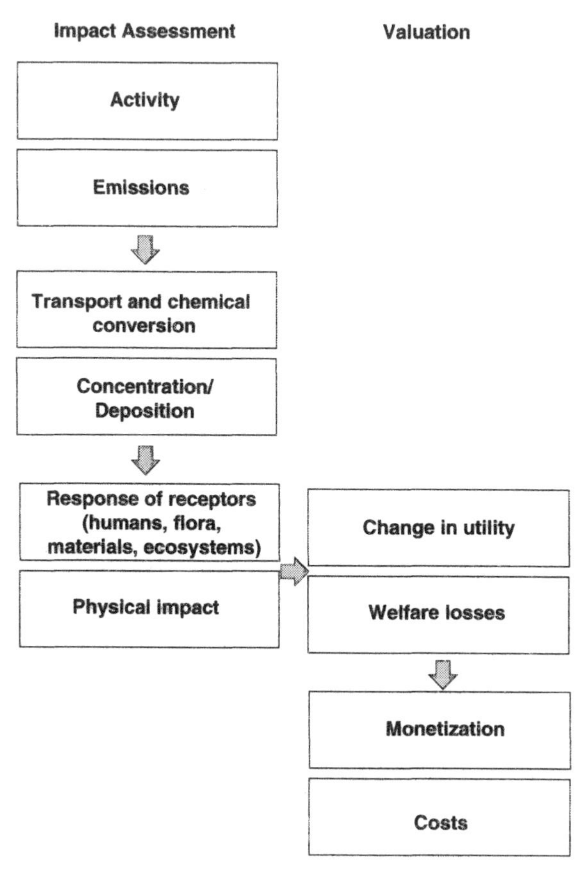

Fig. 1. The Impact Pathway Approach for the Quantification of Marginal External Costs Caused by Air Pollution.

for the different stages. For instance, the number of studies on the relationships between noise exposure and health effects is limited.

Therefore it might be impossible to reliably quantify physical impacts. In this case, relationships linking noise exposure and WTP for reducing amenity losses directly may be applied.

It is important to note, that, although marginal costs are estimated, the pollutant or noise emissions from all other sources and the background burden influence the marginal change due to non-linearities and therefore have to be accounted for in the framework. If emissions of pollutants and noise that occur in the future (or the past) have to be assessed, scenarios of the emissions and concentrations of pollutants at that time have to be used.

The principle of the IPA can be applied to all modes. The character of a burden may differ by mode, as e.g. for noise: roads usually cause a rather constant noise level, while noise from railway lines and airports is characterised by single events with high noise levels. Such differences have to be taken into account and the models used on the respective stage have to be adjusted appropriately. The application of the same approach for all modes ensures consistency of the resulting estimates. Table 3 describes the stages of the general impact pathway for the cost categories.

2.3. Uncertainties and Gaps

The methodology is sometimes criticised by mentioning the uncertainties involved. And indeed uncertainties are large. Before discussing these, one has however to distinguish these uncertainties from deviations of current results compared with earlier results as well from ExternE as from other publications. First there has been a substantial methodological development over the last 15–20 years, e.g. from a top-down to a site-dependent bottom-up approach or with regard to the monetary valuation of health effects. So comparisons should include an analysis of whether the chosen methods are appropriate and state of the art and whether the studies are complete. Second, new knowledge e.g. about health impacts of course changes the results. For example the emerging knowledge that fine particles can cause chronic diseases resulting in a reduction of life expectancy changed the results considerably. An assessment always reflects current knowledge. That an assessment changes with new knowledge – and also may change due to a change in people's preferences – is natural and not a methodological problem.

With regard to uncertainties, first input data and model uncertainties have to be addressed. These can be analysed by using statistical methods. Results show a factor of ca. 2–4 as one-geometric standard deviation intervals around the median estimate (see Rabl & Spadaro, 1999). A factor of 2 would imply that the uncertainty interval ranges from half to double the value reported as best estimate. The largest uncertainties lie in the exposure-response function for health impacts and monetary values for losses in life expectancy. An example with respect to the exposure-response functions is the fact that it is not yet resolved which characteristic of fine particles causes health effects: the number, the mass or the chemical composition. Currently, it is assumed that particles smaller than 1 μm are most important with

Table 3. Description of the General Impact Pathway Per Cost Category.

	Air Pollution	Global Warming	Noise	Soil & Water Pollution	Nuclear Risks
Burden	Emission of airborne pollutants	Emission of greenhouse gases (CO_2, N_2O, CH_4, etc.)	Emission of sound	Emission of solid, liquid and airborne pollutants	Emission of radionuclides in case of an accidental release
Behaviour of burden	Dispersion, chemical transformation, deposition	Amplification of the anthropogenic greenhouse effect, leading to climate change	Propagation	Deposition, infiltration in soil dispersion, chemical transformation in water	Dispersion of nuclear radiation over large areas
Type of impact	Damage to human health, impairment of the natural environment, detrimental effects on man-made environment	Impacts on water supply, coastal areas, agriculture, forestry, energy, sector, species, human health	Damage to human health, amenity losses, impairment of the natural environment	Deterioration and contamination of soil, water resources, impacts on water supply and crops, increase of runoff (inundation) due to sealing	Damage to human health, devastation of soil, crops, water resources and infrastructure in the long term
Receptors affected	Persons, animals, plants, man-made products	Persons, animals, plants, man-made products	Persons, animals	Persons, animals	Persons (living and unborn generation), animals, plants
Monetary values	Market prices if market exists	Market prices if market exists	Market prices if market exists	Market prices if market exists	Market (insurance) prices if market exists
	WTP values if no market price available	WTP values if no market price available	WTP values if no market price available	WTP values if no market price available	WTP values if no market price available

respect to health damages and particle mass is used for measurement, as other measurements are not yet broadly available. Research is directed towards reducing these uncertainties, which reflect our limited knowledge.

Second, certain basic assumptions have to be made like e.g. the discount rate, the valuation of damage in different parts of the world, the treatment of risks with large impacts or the treatment of gaps. Here, a sensitivity analysis should be and is carried out demonstrating the impact of different choices on the results. The full range of results including statistical uncertainties and sensitivities could span a range from a factor of 2 up to 7 around the median value, for global warming impacts even more. Decisions then would sometimes necessitate a choice of the decision maker about the assumption to be used for the decision. This would still lead to a decision process that is transparent and, if the same assumptions were used throughout different decisions, these would be consistent with each other. If due to missing data or understanding of the processes involved it is not possible to make reliable estimates, it may be necessary to use abatement costs as a second best option to obtain quantitative cost estimates. Abatement costs are inferred from reduction targets or constraints for emissions and estimate the opportunity costs of environmentally harmful activities assuming that a specified reduction target is socially desired.

The most prominent example for the need for applying abatement costs instead of damage costs is the area of global warming. Besides missing knowledge about the impacts caused by climate change, uncertainties about scenarios of future developments and about the workings of the climate system are tremendous. Models have been developed to assess the impact of climate change on human health, agriculture, water availability, etc., and quantitative estimates of damage costs have been provided. However, the resulting ranges show up to a factor of 100 between upper and lower bounds. For this reason usually abatement costs are used (e.g. the cost to reach the Kyoto aims for reducing greenhouse gas emissions) to calculate costs from greenhouse gas emissions.

Despite these uncertainties, the use of the methods described here is useful, as

- the knowledge of a possible range of the marginal environmental costs is obviously a better aid for policy decisions than the alternative – having no quantitative information at all;
- the relative importance of different impact pathways is identified (e.g. has benzene in street canyons a higher impact on human health than fine particles?);

- the important parameters or key drivers, that cause high costs, are identified;
- the decision-making process will become more transparent and comprehensible; a rational discussion of the underlying assumptions and political aims is facilitated;
- areas for priority research will be identified.

It is however remarkable that despite these uncertainties certain conclusions and decisions are robust, i.e. do not change over the whole range of possible marginal cost values. Examples are the cost benefit ratio for reducing the fuel sulphur content or coal fired power plant desulphurisation. Furthermore, it should be noted that gaps can be closed and uncertainties reduced by performing further research (e.g. further contingent valuation studies and epidemiological studies).

3. CURRENT RESULTS AND GENERALISATION

The following section gives results of the recent research project UNITE (UNIfication of accounts and marginal costs for Transport Efficiency) and analyses the variability of marginal cost estimates with respect to the question if results can be generalised. This is an important practical issue, because if the generalisation of cost estimates is not possible, specific studies have to be carried out for each new application, which is resource and time consuming. It would be much cheaper if results could be transferred as far as possible to both other vehicle characteristics (e.g. pollutant emissions) and locations.

The cost categories covered in detail are air pollution, noise and global warming. Nuclear risks were included in the UNITE analysis; however, the quantifiable costs proved to be of minor importance compared to air pollution, noise and global warming. Models for estimating impacts from soil and water pollution are currently being developed, though first results are low compared to the other cost categories.

3.1. Costs Due to Air Pollution

The starting point for the bottom-up approach for quantification of marginal costs is the micro level, i.e. the traffic flow on a particular route segment. The marginal costs of one additional vehicle are calculated for a single trip on this route segment, modelling the path from emission to impact and costs. Results of recent bottom-up calculations (see e.g. Bickel et al., 2003; or Friedrich & Bickel, 2001) have shown that the value of externalities may differ substantially from one transport route to another. Two types of emissions are distinguished to account for differences in emission characteristics: direct emissions and indirect emissions.

3.2. Direct Exhaust Emissions

Vehicles belonging to this category have an internal combustion engine. They represent line emission sources, emitting continuously along a route. This group comprises – across modes – conventional road vehicles, diesel locomotives, barges, ships and aircrafts.

3.3. Road Transport

To give an idea of the variation of air pollution costs with vehicle type, fuel and location Table 4 reports costs from road vehicle exhaust emissions quantified in the UNITE case studies for three vehicle technologies: petrol and diesel passenger cars and heavy goods vehicles (HGV), all complying with the EURO2 emission standard. Partly, the costs per vehicle kilometre vary considerably between locations and vehicle/fuel types.

The main parameters determining the costs due to direct vehicle emissions are:

- Emission factors, which differ by fuel (e.g. petrol–diesel), vehicle type (e.g. heavy diesel vehicles–diesel cars), emission standard (e.g. EURO2 versus EURO4), and driving pattern (speed, acceleration processes).
- The local environment close to the road (receptor density, meteorology, above all average wind speed up to ca. 25 km from the road – later referred to as *local scale*).
- The geographical location within Europe (determining the number of receptors affected by long-range pollutant dispersion and formation of secondary pollutants, which is modelled for Europe as a whole – later referred to as *regional scale*).

To eliminate the effect of different emission factors used in the case studies and to allow the analysis of differences between locations, damage costs are related to a unit of emission of $PM_{2.5}$. $PM_{2.5}$ is one of the key pollutants as

Table 4. Overview of Damage Costs Due to Air Pollution from Road Vehicle Exhaust Emissions in €cent/vkm.

		Car Petrol EURO2	Car Diesel EURO2	HGV Diesel EURO2
Urban case studies	Helsinki	0.12	n.a.	n.a.
	Stuttgart	0.25	1.45	17.52
	Berlin	0.15	0.73	10.19
	Florence[a]	*0.01*[a]	*0.26*[a]	*4.69*[a]
Inter-urban case studies	Helsinki–Turku	n.a.	n.a.	2.09
	Basel–Karlsruhe	0.37	0.63	6.91
	Strasburg–Neubrandenburg (outside built-up areas)	0.12	0.26	3.89
	Strasburg–Neubrandenburg (within built-up areas)	0.11	0.38	7.46
	Milano–Chiasso	0.25[b]	1.91[b]	6.72[b]
	Bologna–Brennero	0.20[b]	0.73[b]	5.07[b]

Source: Bickel et al. (2003).
[a]Restricted comparability to other results, because estimate is based on a different methodological approach; only human health impacts on the local scale due to CO, Benzene and PM_{10} considered (NO_x, SO_2, Ozone, NMVOC not included).
[b]Emission standard not specified.

regards the share in the costs quantified, in particular on the local scale (i.e. up to ca. 25 km from the emission source). Table 5 shows damage costs on the local and on the regional scale (covering long-range pollutant transport all over geographical Europe). Costs on both scales add up to the total costs caused by a unit of pollutant emitted in the respective area. In urban areas, local scale effects dominate the costs due to the high receptor density. For this reason, it is one of the hypotheses that costs can be generalised based on the number of population affected in the urban area.

But a closer look at the case study results reveals, that there must be another relevant parameter. In Helsinki and Stuttgart, the number of persons affected is almost the same, in absolute and in relative (inhabitants per km^2) terms. Even though, in Stuttgart costs on the local scale are about a factor of two higher than in Helsinki. Furthermore, in Berlin where many more persons are affected than in Helsinki and Stuttgart, the local scale costs per unit of $PM_{2.5}$ emitted are lower than in the other two cities. These differences in the local scale costs are determined by the local meteorology. In Stuttgart, the yearly average wind speed is much lower than for example in Berlin, so that pollutants stay longer in the densely populated area before being dispersed, resulting in higher damages. As a consequence, local

Table 5. Comparison of Damage Costs Due to $PM_{2.5}$ Emissions of Urban Case Studies.

Location	Population Density (in h/ km^2)	Costs Due to Damages on the Local Scale in €/tonne of $PM_{2.5}$	Costs Due to Damages on the Regional Scale in €/tonne of $PM_{2.5}$
Helsinki	2,800	95,000	2,800
Stuttgart	2,800	200,000	26,800
Berlin	3,800	90,000	17,500
Florence[a]	*4,100*	*50,000*[a,b]	n.a.

Source: Bickel et al. (2003).
[a]Restricted comparability to other results, because estimate is based on a different methodological approach.
[b]€/t PM_{10}; n.a. = not available.

meteorology has to be taken into account when generalising cost estimates for emissions in urban areas.

Regional scale costs depend on the geographical location within Europe, determining the number of receptors affected and when looking at other pollutants than $PM_{2.5}$ the formation of secondary pollutants (above all ozone as well as nitrate and sulphate aerosols) via air chemistry. Such differences may be considerable as illustrated in Table 5.

Due to the prevailing west winds, emissions in the south of Finland are transported towards the Baltic sea and to sparsely populated areas in Russia, leading to very low costs per unit of pollutant emitted. Even within Germany, there are considerable differences between emissions in the (North-) East and the Southwest, because the areas affected by emissions in the Southwest are more densely populated. Besides, the formation of sulphates and nitrates is influenced by the background concentrations of the reactive species involved and the ratio of SO_2 and NO_x emitted.

Whereas for urban areas the share of local scale costs is high and therefore the geographical location is of minor importance, the regional scale damages are very important for locations outside urban areas.

As a consequence the generalisation of results has to take into account at least the parameters

(a) for emissions in urban areas: the local population density and the meteorology;
(b) for emissions outside of urban areas: the geographical location and the character of the route (passing built-up areas or crossing built-up areas).

3.4. Other Modes

The relationships between pollutant emission and associated costs are in principle the same as for road transport. Emissions from diesel locomotives, maritime and inland waterway vessels can be treated like emissions from road vehicles, taking into account the character of the route. For these modes, however, the main part of emissions will occur outside the urban areas.

The costs due to emissions of a typical passenger ferry travelling from Helsinki (Finland) to Tallinn (Estonia) were estimated as about €12 per vessel kilometre at open sea. The costs for a container barge between Rotterdam (Netherlands) and the inland port of Mannheim (Germany) amount to €3.60 upstream and €2.40 downstream (including global warming).

Aircraft emissions are a special case, because most of the emissions take place in high altitudes. Assessment of the resulting impacts is still to be improved, because modelling of dispersion and chemical conversion is not as advanced as for low-level emissions. On the other hand impacts due to low-level emissions at airports occurring during arrival, ground activities and departure can be assessed with the existing models. For a flight of a Boeing 737-400 from Berlin to London air pollution costs of about 5 €cent per aircraft kilometre were estimated.

3.5. Indirect Emissions from Electricity Production and Fuel Production

Damages and costs due to air pollution from electricity production depend on where the electricity is produced and thus where pollutant emissions from power plants occur. Based on the emissions for producing a unit of electrical energy (e.g. 1 kWh), damage costs per kWh can be calculated. If electricity is generated from different fuels, the resulting costs per kWh can be combined according to the share of the different fuels in the electricity production. Table 6 illustrates the fuel mixes considered in the UNITE case studies for Germany. The resulting costs per kWh of electricity produced vary considerably, mainly depending on the share of fossil fuels used: by a factor of three between Stuttgart urban transport and the German Rail and a factor of four between Stuttgart and Berlin.

Generalisation of costs per kWh of electricity requires information on the geographical location of the power plant and the respective emission factors. If the electricity is produced from different fuels, the costs per kWh can be calculated according to the share of the different fuels in the electricity production.

Table 6. Share of Fuels in the Electricity Production of Rail Transport Operators.

	Berlin	Stuttgart	German Rail
Coal (%)	74.0[a]	1.9[b]	34.4[c]
Nuclear (%)		89.6[b]	22.1[c]
Oil/Natural gas (%)	26.0[a]	1.9[b]	13.2[c]
Hydro (%)		2.1[b]	10.1[c]
Electricity from public grid and other (%)		4.5[b]	20.2[c]
Total (%)	100.0	100.0	100.0
Environmental marginal cost (€cent per kWh)	2.10	0.49	1.53

[a]BEWAG (2002).
[b]Schmid, Wacker, Kürbis, Krewitt, and Friedrich (2001).
[c]Deutsche Bahn (1998).

Costs due to indirect emissions from fuel production gain in importance for vehicles complying with stricter emission standards. For petrol cars complying with EURO2 standard or higher, the costs from fuel production may reach the same order of magnitude as the costs from exhaust emissions. For diesel vehicles this is usually not the case, because the costs due to exhaust emissions are generally higher, and the costs due to diesel production are lower than for producing petrol.

It was beyond the scope of the UNITE case studies to take into account where the fuel burnt specifically was produced. Therefore, average costs for the emissions due to fuel production within a country were quantified, taking into account emission factors for fuel production in refineries and the specific costs due to air pollution from the refineries within a country. These values only vary with the fuel consumption of a vehicle and therefore can be generalised – based on the fuel consumption – for the country for which they were calculated.

3.6. Generalisation Aspects

A comparison of the case study results clearly suggests that a direct transfer of costs due to air pollution cannot be recommended. Some general rules could be derived, but an operational formula for transfer requires a broader statistical basis of case studies. A generalisation methodology for air pollution costs should account for

- the local scale conditions (population density and local meteorology), and
- the regional scale costs per tonne of pollutant emitted in a certain area (e.g. on NUTS1 level)

Table 7 presents a list of the main generalisation aspects.

3.7. Costs Due to Noise

Noise is a very local burden, marginal costs due to noise exposure are mainly determined by

- the distribution and distance of exposed persons from the source;
- the existing noise level, which in most cases is dominated by the traffic (number of vehicles, trains or aircraft per hour, mix of vehicle, train or aircraft types, speed);
- the time of day (variation in disturbance effects of noise).

Table 8 illustrates the broad variation in noise costs quantified in the UNITE case studies. Marginal costs are generally higher at nighttime than during daytime, with a difference of up to a factor of 3. One reason for this is the higher disturbance effects of noise during the night. The second reason is a lower background noise level at night. Due to the logarithmic nature of noise the effect of an additional noise emission decreases with increasing background noise. In other words: the higher the background noise level the lower the noise costs of an additional vehicle. Differences between the case studies are large, reflecting the variability of marginal costs with the detailed population distribution, number and speed of vehicles, share of HGVs, etc. For example in Berlin the average number of persons per road kilometre affected by noise is slightly higher than in Stuttgart. However, the costs are more than a factor of three lower due to the much higher number of vehicles and higher speeds on the street considered in the case study. The comparably low costs for Helsinki are caused by the lower number of people affected on the route considered. The same is the case when comparing the results of the urban case studies to the inter-urban case studies, which mainly run outside built-up areas.

The methodology can be generalised, considering mode-specific characteristics of noise propagation and exposure-response functions. A generalisation of input data and results is very difficult, due to the non-linearities involved and the variability of the local characteristics. Table 9 presents the main generalisation aspects.

Table 7. List of Generalisation Aspects for Marginal Costs Due to Air Pollution.

	Aspect to be Generalised	Important Mainly for	Basic Requirements for Generalisation
Methodology	Overall methodology: IPA		Can be generalised
Inputs	Inputs to dispersion models		Generalisation not recommended
	Exposure–response functions		Can be generalised
	Monetary values for health impacts		Country-specific adjustment/values for local scale impacts
	Exhaust emission factors for specific vehicle technologies		Same emission standard; same driving characteristics/speeds
	Exhaust emission factors for vehicle fleets		Generalisation not recommended
	Emission factors for the production and transport of fuel		Refinery processes and fuel distribution are comparable
Output values	Regional scale unit costs per tonne of pollutant	Emissions outside of urban areas	Pollutant is emitted in the same geographical area (e.g. administrative unit on NUTS1 level)
	Local scale unit costs per tonne of pollutant for low-level emissions from vehicles with internal combustion engine	Urban emissions	Comparable local environment, i.e. population density and local meteorology; country-specific adjustment of monetary values
	Costs due to fuel production (per litre of fuel or per vehicle kilometre)	Emissions outside of urban areas	Comparable emission factors for production and transport of fuel; pollutants are emitted in the same geographical area
	Costs from exhaust emissions per vehicle kilometer	Both urban emissions and emissions outside of urban areas	Comparable emission factors and local environment and geographical area/ regional scale unit costs

Table 8. Overview of Costs Due to Noise from Road Vehicles in €cent/vkm.

		Passenger Car		HGV	
		Daytime	Nighttime	Daytime	Nighttime
Urban case studies	Helsinki	0.22	0.53	n.a.	n.a.
	Stuttgart	1.50	4.50	25.75	78.25
	Berlin	0.47	1.45	7.67	23.33
	Florence[a]	n.a.	n.a.	n.a.	n.a.
Inter-urban case studies	Helsinki–Turku	n.a.	n.a.	1.58	3.86
	Basel–Karlsruhe	0.02	0.03	0.11	0.18
	Strasburg–Neubrandenburg (outside built-up areas)	0	0	0	0
	Strasburg–Neubrandenburg (within built-up areas)	0.12	0.19	3.04	5.06
	Milano–Chiasso	0.01	0.04	0.09	0.35
	Bologna–Brennero	0.001	0.002	0.006	0.02

Source: Bickel et al. (2003).
[a]Marginal costs not given per vehicle kilometre, but per 1 dB(A) reduction.

Marginal noise costs due to maritime shipping and inland waterway transport were assumed to be negligible, because emission factors are comparably low and most of the activities occur outside densely populated areas and therefore relevant thresholds for observing effects are expected to be not exceeded.

3.8. Costs Due to Global Warming

Damages due to climate change is one of the most important categories of fossil fuel emission related damages, but also among the most uncertain and controversial. The impacts of incremental climate change due to emissions from the European transport sector are global in range. Moreover, they will extend over many generations and a very wide range of resources and human activities. The Inter-Governmental Panel on Climate Change (IPCC)

Table 9. List of Generalisation Aspects for Marginal Costs Due to Noise.

	Aspect to be Generalised	Basic Requirements for Generalisation
Methodology	Overall methodology: Impact Pathway Approach	Can be generalised
Inputs	Inputs to noise propagation models (different models for road, rail and aircraft transport required)	Generalisation not recommended
	Exposure-response functions for health impacts (partly mode-specific)	Can be generalised (for same mode)
	Monetary values for health impacts and amenity losses	Country-specific adjustment/values
	Noise emission factors for vehicle/train/aircraft types	Only if same parameters (e.g. driving characteristics, speeds)
	Noise emission factors for vehicle fleets	Generalisation not recommended
Output values	Costs per vehicle/train kilometre or aircraft movement	Generalisation difficult due to high sensitivity concerning local characteristics

has undertaken detailed analyses of the science of climate change, the physical impacts and potential responses and the socio-economic dimensions. Climate damage models consider effects of a climate change on agriculture, health damages and benefits, energy use, water availability, etc. on different scales. However, the results from such models have to be interpreted very cautiously and, as mentioned above (see Section 2.3), it is advisable to use abatement costs for valuation.

Due to the global character of global warming impacts, there is no difference where the emission of greenhouse gases takes place. A European perspective was taken for deriving the abatement cost value, i.e. the value is applicable to all countries of the European Union. It is based on calculations for reaching the Kyoto targets of the European Union, optimising abatement measures in the Union and not country by country. It is assumed that measures for a reduction in CO_2 emissions are taken in a cost-effective way, implying that reduction targets are not set per sector, but that the cheapest measures are implemented, no matter in which sector or country. Based on estimates by Capros and Mantzos (2000) a value of €20 per tonne of CO_2 emitted was used. This value represents a central estimate of the range of values for meeting the Kyoto targets in 2010 in the EU. They report a value of €5 per tonne of CO_2 avoided for reaching the Kyoto targets for the EU, assuming a full trade flexibility scheme involving all regions of the world. For the case that no trading of CO_2 emissions with countries outside the EU is permitted, they calculate a value of €38 per tonne of CO_2 avoided. The value of €20 is in line with results from other studies, for instance Fahl, Läge, Remme, and Schaumann (1999) estimate €19 per tonne of CO_2 for meeting a 25% emission reduction from 1990 to 2010 in Germany.

Even though the costs presented in Table 10 are calculated using the same value for all case studies, there are variations. These illustrate the differences in the emission factor per kilometer, which mainly stem from different fuels used (Petrol versus Diesel) and operation parameters (e.g. traffic situation).

The approach for calculating costs from greenhouse gas emissions is very simple. Due to the global scale of the damage caused, there is no difference how and where the emissions take place (as long as we do not talk about aircraft emissions at cruising altitude). The amount of greenhouse gas emitted is multiplied by a specific cost factor. This procedure is transferable per se as no other parameters have to be taken into account. Table 11 lists the main generalisation aspects.

Table 10. Overview of Damage Costs Due to Global Warming from Exhaust Greenhouse Gas Emissions from Road Vehicles in €cent/vkm.

		Car Petrol EURO2	Car Diesel EURO2	HGV Diesel EURO2
Urban case studies	Helsinki	0.35	n.a.	n.a.
	Stuttgart	0.47	0.31	3.28
	Berlin	0.47	0.31	3.28
	Helsinki–Turku	n.a.	n.a.	2.40
	Basel–Karlsruhe	0.37	0.32	2.18
Inter-urban case studies	Strasburg–Neubrandenburg (outside built-up areas)	0.34	0.25	2.03
	Strasburg–Neubrandenburg (within built-up areas)	0.47	0.31	3.28
	Bologna–Brennero	0.36[a]	0.36[a]	2.16[a]

Source: Bickel et al. (2003).

[a]Emission standard not specified.

Table 11. List of Generalisation Aspects for Marginal Costs Due to Global Warming.

	Aspect to be Generalised	Basic Requirements for Generalisation
Methodology	Overall methodology: application of abatement costs	Can be generalised
Inputs	Monetary value per unit of greenhouse gas emitted	Value used for calculations is applicable for countries of the European Union
	Exhaust emission factors for specific vehicle/train/vessel/aircraft technologies	Only if same parameters (e.g. driving characteristics, speeds)
	Exhaust emission factors for vehicle fleets	Generalisation not recommended
	Emission factors for the production and transport of fuel	Refinery processes and fuel distribution are comparable
Output values	Costs from exhaust emissions per vehicle kilometer	Comparable emission factors
	Costs due to fuel production (per litre of fuel or per vehicle kilometre)	Comparable emission factors for production and transport of fuel

4. CONCLUSIONS

Approaches for quantifying marginal costs due to air pollution, noise and global warming are available for all modes. The impact pathway approach represents the preferable method, because it is the only approach capable of quantifying site- and technology-specific marginal costs. In the following, conclusions are drawn with respect to methodology, policy and research.

4.1. Methodology

The IPA, including the respective dispersion models, exposure–response functions and monetary values, is well established and has been applied in a large number of research projects. It is equally applicable to each mode and location, even if dispersion models may have to be modified to reflect mode-specific characteristics. This is the case in particular for noise propagation modelling for road, for rail and for aircraft transport.

For air pollution, the local meteorology was found to be a very important determinant for air pollution costs in urban areas. For this reason special attention should be paid to the quality of this kind of data in future calculations.

The application of the IPA in the context of noise is relatively new and may be subject to revision and extension in the future, in particular the exposure–response functions. Due to the very local nature of noise and the strong non-linearities involved, a lot of detailed input data is required. Availability of noise emission factors for different vehicles is only limited.

In the context of estimating costs due to global warming the main source of uncertainty is the valuation factor used. This reflects the huge uncertainties about the share of anthropogenic emissions in climate change and the associated effects, as well as reasonable reduction targets for greenhouse gas emissions. Compared to this, the emission factors for CO_2 and other greenhouse gases appear very reliable.

4.2. Policy

Analysis of the available estimates show that due to the significant variations of results between locations, reflecting the different characters and conditions of the relations, it is difficult to draw general conclusions concerning magnitude and composition of costs. It is not possible to derive one single value for the marginal environmental costs of a certain vehicle type in urban areas. Therefore, the cost categories have to be looked at separately, in particular when it comes to the issue of generalisation of results.

Quantifiable air pollution costs are dominated by health effects, in particular loss of life expectancy. Diesel vehicles cause considerably higher costs than petrol vehicles, resulting from much higher emissions of primary particles. The difference between both fuel types is highest in urban areas with a diesel car causing about five times higher costs than a petrol car (both complying with emission standard EURO2), because primary particles have very high local effects.

Besides exhaust emissions, emissions due to fuel production processes are relevant and gain in importance with stricter emission standards for road vehicles. For petrol cars complying with the EURO4 standard, costs due to fuel production are comparable to those from exhaust emissions. In the case of electric trains, the marginal costs quantified vary heavily depending on the fuel mix from which the electricity is produced – the lower the share of fossil fuels, the lower the resulting costs.

Further to variations in emission factors the costs are determined by the population density close to the emission source, the local meteorology (mainly average wind speed) and by the geographical location within Europe, which is important for the number of the population affected by long-range pollutant transport and the formation of secondary pollutants.

Noise costs are extremely variable; nighttime values up to a factor of three higher than daytime values can be observed for road transport vehicles in urban areas. Outside of urban areas, absolute levels and differences between daytime and nighttime are much smaller.

Noise costs are mainly determined by the population affected, the time of day (with higher disturbance effects of noise during the night) as well as the number of vehicles and their speeds, and the resulting background noise. The higher the existing background noise level, the lower the costs of an additional vehicle.

The possible effect of marginal cost pricing in the case of noise is a very good illustration of issues that have to be taken into account in the case of strong non-linearities of impacts. A pricing scheme based on marginal noise costs would lead to a bundling effect of traffic. Marginal costs are strongly decreasing with increasing background noise levels. For this reason driving on a route where noise levels are high already would be much cheaper than driving in quiet areas. Of course this is a perfect solution for allocation of noise emitters from the perspective of economic theory. However, it has to be ensured that no absolute limits, such as thresholds for health risks or amenity losses are exceeded. In practice, other price components (e.g. air pollution costs) may attenuate the bundling effect.

For petrol cars complying with EURO2 emission standard, marginal costs associated to greenhouse gas emissions are higher than the costs due to air pollution; the same holds for the aircraft studied. Due to the higher emission of primary particles, for diesel engines global warming costs are generally lower than air pollution costs.

4.3. Research

In general, the models and data available reflect best current knowledge; however, this knowledge has gaps and therefore the results are subject to uncertainty. Further research is needed to reduce the uncertainties.

Compared to the pollutant emission factors available for road vehicles, those available for marine and inland waterway vessels as well as for aircraft are less elaborated. In the special case of high-altitude aircraft emissions information on the dispersion and chemical conversion of pollutants is urgently required to be able to assess the impacts of all flight phases adequately.

In the context of generalisation of results the comparison of results clearly suggests that costs due to air pollution cannot be transferred based on the population density of the local environment only. Some general rules can be derived, but an operational formula for transfer requires a broad statistical basis of case studies. A generalisation methodology for air pollution costs should account for

- the local scale conditions (population density and local meteorology), and
- the regional scale costs per tonne of pollutant emitted in a certain area (e.g. NUTS1 level)

For marginal noise costs, a generalisation is even more difficult than for air pollution, because of the large non-linearities involved and the variability of the relevant parameters in very short time spans. The availability of noise emission factors for different vehicles should be improved.

Costs due to the emission of greenhouse gases are not location specific, as they are relevant on a global scale, therefore generalisation is not a problem.

REFERENCES

AEA Technology. (1997). *Economic evaluation of air quality targets for CO and benzene.* Interim report prepared for the European Commission DG XI, Harwell, UK.

BEWAG. (2002). *Kraftwerkstypen.* www.bewag.de/_energie/content7106.asp

Bickel, P., Schmid, S., Tervonen, J., Hämekoski, K., Otterström, T., Anton, P., Enei, R., Leone, G., van Donselaar, P., & Carmigchelt, H. (2003). *Environmental marginal cost case studies.* UNITE Deliverable 11, Stuttgart. http://www.its.leeds.ac.uk/projects/unite/D11.pdf

Capros, P., & Mantzos, L. (2000). Kyoto and technology at the European Union: Costs of emission reduction under flexibility mechanisms and technology progress. *International Journal of Global Energy Issues, 14,* 169–183.

Cline, W. R. (1992). *The economics of global warming.* Washington, DC: Institute for International Economics.

Deutsche Bahn. (1998). *Umweltbericht 1997.* Berlin.

ECMT (European Conference of Ministers of Transport). (1998). *Efficient Transport for Europe: Policies for Internalisation of External costs.* Paris.

European Commission. (1998). *Transport research – Fourth framework programme.* QUITS: Quality Indicators for Transports Systems, Office of Publications for the European Communities, Luxembourg.

European Commission. (1999a). *ExternE externalities of energy. Vol. 7 – Methodology 1998 update.* A report produced for the EC – DG XII, Luxembourg, Office of Publications for the European Communities, Luxembourg.

European Commission. (1999b). *ExternE externalities of energy. Vol. 10 – National implementation.* A report produced for the EC – DG XII, Luxembourg, Office of Publications for the European Communities, Luxembourg.

Fahl, U., Läge, E., Remme, U., & Schaumann, P. (1999). *E3Net. In: Forum für Energiemodelle und Energiewirtschaftliche Systemanalysen in Deutschland* (Hrsg.) Energiemodelle zum Klimaschutz in Deutschland. Heidelberg: Physica-Verlag.

Fankhauser, S. (1995). *Valuing climate change – The economics of the greenhouse.* London: EarthScan.

Friedrich, R., & Bickel, P. (Eds) (2001). *Environmental external costs of transport.* Heidelberg: Springer-Verlag.

INFRAS/IWW. (2000). *External costs of transport.* Study for the International Union of Railways, Paris.

Link, H., Stewart, L. (DIW), Doll, C. (IWW), Bickel, P., Schmid, S., Friedrich, R., Krüger, R., Droste-Franke, B., & Krewitt, W. (IER). (2001). The Pilot Accounts for Germany. UNITE Deliverable 5 Annex 1, Leeds, UK.

McCubbin, D. R., & Delucchi, M. A. (1996). *The social cost of the health effects of motor-vehicle air pollution.* Report #11 in the Series: The Annualised Social Cost of Motor-Vehicle Use in the United States, based on 1990-1991 Data, UCD-IST-RR-96-3 (11), Davis, California, US.

Nordhaus, W. D. (1991). To slow or not to slow: The economics of the greenhouse effect. *Economic Journal, 101*, 920–937.

Rabl, A., & Spadaro, J. (1999). Environmental damages and costs: An analysis of Uncertainties. *Environmental International, 25*(1), 29–46.

Schmid, V., Wacker, M., Kürbis, I., Krewitt, W., & Friedrich, R. (2001). *Systematischer Vergleich konkreter Fahrten im Personenverkehr im Hinblick auf umwelt- und klimarelevante Wirkungen verschiedener Verkehrsmittel.* Projekt im Rahmen des Förderprogrammes BWPlus, Förderkennzeichen PEF 498 001, Institut für Straßen- und Verkehrswesen und Institut für Energiewirtschaft und rationelle Energieanwendung, Universität Stuttgart, Stuttgart.

Titus, J. G. (1992). The costs of climate change to the United States. In: S. K. Majumdar, L. S. Kalkstein, B. M. Yarnal, E. W. Miller & L. M. Rosenfeld (Eds), *Global climate change: Implications, challenges and mitigation measures* (pp. 384–409). Easton: Pennsylvania Academy of Science.

Tol, R. S. J. (2001). Global Warming. In: R. Friedrich & P. Bickel (Eds), *Environmental external costs of transport* (pp. 384–409). Heidelberg: Springer-Verlag.

World Health Organisation (WHO). (1999). *Health costs due to road traffic – related air pollution – An impact assessment project of Austria, France and Switzerland.* Synthesis report, Federal Department of Environment, Transport, Energy and Communications – Bureau for Transport Studies, Bern, Switzerland.

THE IMPACTS OF MARGINAL SOCIAL COST PRICING

Inge Mayeres, Stef Proost and Kurt Van Dender

1. INTRODUCTION

While the previous chapters have dealt with the measurement of the marginal social costs of transport, the aim of this chapter[1] is to explore the impacts of marginal social cost (MSC) pricing. In the economic literature, short run MSC pricing is the benchmark for an efficient transport pricing policy. This principle holds when there are no other price distortions in the economy and no income distribution concerns. The short run MSC contains the marginal resource costs,[2] marginal infrastructure costs, scarcity or external congestion costs and external environmental and accident costs. The short run MSC principle ensures that the existing infrastructure is used as efficiently as possible. Whether the infrastructure is at an efficient level or not and whether it is optimally maintained or not has an impact on the level of the MSC but not on the short run MSC pricing rule itself.

Apart from pricing measures, policy makers also use investment as a policy instrument. To determine an efficient investment level requires a cost–benefit analysis that trades off the benefits of infrastructure extension (discounted sum of saved user costs, including time, saved external accident and environmental costs, reduced maintenance costs) and the costs of an infrastructure extension (investment costs). This rule only holds if there is short run MSC pricing and if there are no other distortions in the economy.[3]

The pricing and investment rules get more complicated in the presence of budget constraints or when there are restrictions on the available policy

Measuring the Marginal Social Cost of Transport
Research in Transportation Economics, Volume 14, 211–243
Copyright © 2005 by Elsevier Ltd.
All rights of reproduction in any form reserved
ISSN: 0739-8859/doi:10.1016/S0739-8859(05)14008-6

instruments (e.g. when prices cannot be tailored to each transport market separately). In those cases short run MSC pricing is no longer optimal. However, the resulting second-best pricing rules[4] are based on carefully balanced deviations from the short run MSC. This implies that MSC information remains crucial. It also implies that average cost (AC) pricing, which is based on average costs rather than marginal social costs, is an inefficient way of achieving balanced budgets.

Which budget constraints make sense for the transport sector? Budget constraints are motivated because they are meaningful ways to limit the use of general tax revenues by the transport sector. Collecting tax revenue in other sectors is costly too because these taxes create economic distortions. The two types of inefficiencies (pricing above short run MSC in the transport sector versus creating a wedge between marginal social costs and prices in other sectors) need to be balanced. This can be done using a shadow cost of public funds that should ideally be computed at the level of the economy, and whose value depends on which tax instrument is used to recycle the net tax revenues from the transport sector. In the absence of equity concerns, balancing inefficiencies in the transport sector with the rest of the economy also implies that extra tax revenues collected in the transport sector should be used to reduce existing tax inefficiencies in the rest of the economy in the form of lower labour or capital taxes.

Including income distribution concerns in transport pricing policies is a tough problem. Income distribution concerns can be best met by considering a wider set of instruments than transport prices and subsidies. There are three important criteria for the appropriateness of subsidies or lower taxes on a given good. First, there is the price elasticity of the good in question. From an efficiency point of view it is good to impose a higher tax on goods with a lower price elasticity in order to limit the resource allocation losses of distorted prices. Second, there is the relative consumption of the good by low-income classes. Third, there is the way the subsidy or lower tax is financed: who pays the higher taxes on other goods will depend again on who consumes these other goods and the price elasticity of these other goods matters to keep efficiency losses down. This requires analysing the effects in the rest of the economy. Therefore, a general equilibrium assessment would be ideal.

The previous chapters described how to derive the marginal social costs of transport. The objective of this chapter is to give a better insight in the impacts of MSC pricing on prices, traffic levels, revenues, welfare and income distribution. In addition, the chapter compares MSC pricing with an alternative pricing rule: AC pricing. We also explore the impacts of the pricing reforms on the transport accounts. These make a comparison between the costs of transport and the revenues from transport pricing, charging and taxation (Link et al., 2002). We test whether the changes in the transport accounts that are recorded after the pricing reforms are good welfare indicators. Section 2 describes in more detail what these pricing rules entail and presents the basic features of the two models that are used to calculate their impacts. The first model is a partial equilibrium model, while the second model is a computable general equilibrium model.

Section 3 reports the results of the partial equilibrium case studies. The main objective is to illustrate the direct effects of alternative transport pricing approaches. The direct effects are the effects on transport prices and tax levels, the effects on transport volumes and the effect on welfare. Section 4 presents the general equilibrium results that give a better insight in the indirect effects of the pricing reforms. The focus lies on the specific effects of the way in which transport sector deficits are financed by other taxes and on the equity impacts. Section 5 formulates the main conclusions that can be drawn from the case studies.

2. THE BASIC FEATURES OF THE SIMULATION EXERCISES

2.1. The Pricing Scenarios

In this chapter we test two archetypes of pricing rules: marginal social cost (MSC) pricing and average cost (AC) pricing.

Marginal Social Cost pricing means that prices are set equal to sum of the marginal resource cost, marginal infrastructure cost and the marginal external cost, all this for a given infrastructure. The term "marginal cost" refers to the cost caused by an additional passenger kilometre (pkm) or vehicle kilometre (vkm). Under the MSC pricing principle there is no consideration whatsoever of the financial impact per mode. It is also assumed that there are no implementation costs.

Average Cost Pricing is defined as follows: prices for a certain mode are set equal to the sum of financial costs of that mode divided by the total volume of that mode. This implies that no attention is paid to the structure of the resource costs (fixed or not, sunk or not, etc.), that there is no consideration of the external costs[5] and that all transport services (freight, passengers, etc.) within that mode are treated identically. The main goal of AC pricing is financial cost recovery.

2.2. The Models: Partial versus General Equilibrium Approach

We test the effects of these pricing rules using two types of models: a partial equilibrium model and a general equilibrium model. Table 1 compares their main features.

The partial equilibrium analysis is carried out with the TRENEN model (Proost and Van Dender, 2001a,b). Compared to the general equilibrium model it can analyse in a much more detailed way the transport markets and different pricing policies. It generates detailed effects on transport volumes and on the efficiency of the transport sector. Results will be presented for London and the UK southeast region. The general equilibrium model is a computable general equilibrium model for Belgium (Mayeres, 1999, 2000). It cannot offer the same degree of modelling detail of the transport sector. However, it offers other important advantages. Firstly, it allows to model

Table 1. The Features of the Models Used.

	Partial Equilibrium Model	General Equilibrium Model
Focus	Transport sector of a region or a country (different modes)	Whole economy of a country
Markets modelled	All transport markets of a region or country	Transport markets, labour market and other input markets, markets of all consumption goods
Cost or benefit of extra tax revenues raised in the transport sector	Exogenous – here put equal to 1[a]	Endogenous, will depend on the way the extra revenue is used
Equity issues	Are not dealt with because the use of the surplus or financing of the deficit is not specified	Studied by income group
Welfare measure	Sum of consumer and producer surplus, tax revenues and external costs on the transport markets	Differences in utility for different households
Model case studies	Two regions in UK (Greater London and the southeast region)	Belgium
Infrastructure	Fixed	Fixed

[a]The partial equilibrium model assumes a first best economy: perfect lump sum redistribution of revenues is possible and there are no distortions in the rest of the economy.

the economic costs of financing a larger deficit in the transport sector. Any increase (decrease) in the deficit (surplus) in the transport sector will require an increase of labour or other taxes and this may be more or less costly. The second advantage of the general equilibrium model is that it allows to track better the full incidence of tax reforms on the utility of different individuals. It is therefore better suited for an analysis of the equity effects. In both the partial and the general equilibrium model the pricing reforms are evaluated for a given infrastructure.

3. THE DIRECT EFFECTS OF MARGINAL SOCIAL COST PRICING AND AVERAGE COST PRICING

This section illustrates the direct effects of MSC pricing and compares them with the direct effects of AC pricing.

We present results for two regions in the UK: Greater London and the southeast region. The two pricing rules are compared in terms of their effect on social welfare. Since this section disregards income distribution issues and effects on other sectors, social welfare equals the unweighted sum over all transport markets of consumer surplus, producer surplus, external costs and government revenues. Note that even if the principle of social welfare maximisation is rejected as a basis for policy design because of government failures, it is still a useful reference point for the evaluation of different policy approaches.

Section 3.1 gives a brief overview of the TRENEN model that is used to calculate the direct effects of the pricing rules. Section 3.2 discusses the implementation of the pricing scenarios in TRENEN. The main results are summarised in Section 3.3.

3.1. The TRENEN Model in Brief

The TRENEN model is a partial equilibrium, multi-modal representation of urban and interregional transport systems. It was developed within the EC 4th Framework Programme. For a detailed description, the reader is referred to Proost and Van Dender (2001a,b). The model computes the efficiency effects of pricing policies, either using exogenously specified prices (simulation), or finding the optimal price levels by maximising a social welfare function (optimisation). In the latter case, pricing constraints can be taken into account. In the present version of the model the transport network infrastructure (for road, rail, metro, water) is kept fixed in all case studies.

The model contains a detailed breakdown of transport markets by transport mode, time of day (peak and off-peak), and environmental characteristics of vehicles. However, area and route type (urban/interurban/rural) are not included. Typically, some 20 markets are distinguished in a case study. The degree of detail in modelling the transport markets is important for assessing the potential of alternative pricing instruments. The more disaggregate the structure, the higher the benefits one can expect from MSC pricing relative to single aggregate AC pricing. The passenger transport markets are represented by a utility tree. The representative consumers' utility is defined over a composite non-transport commodity and over an aggregate transport commodity. The breakdown of the latter is similar across different case studies, while details may differ in order to correspond with the particular context. The structure of the utility tree reflects assumptions on the substitutability between transport markets. Moving further down the tree generally implies easier substitution. Parameters are chosen so that the implicit price elasticities of demand correspond to values found in the literature.

Demand for freight transport is generated by a representative producer's cost function. To produce a composite commodity, freight and other inputs are required. In a typical case study, a choice can be made between peak and off-peak freight, and in both periods different modes (rail, road, inland waterways) are available. Road freight can make use of light-duty and heavy-duty trucks.

The main transport externalities (congestion, air pollution, accidents and noise) are taken into account. Marginal external congestion costs depend on traffic volumes, while the other external costs are taken to be constant per vkm in each transport market.[6]

The cost structure of transport activities in TRENEN is simple. Marginal resource costs of inputs other than time are constant per vkm, within each transport market. They may differ across transport markets (i.e. across modes, time periods, vehicle types, etc.). Furthermore, in passenger car markets no fixed costs are present.[7] Average resource costs of non-time inputs then are equal to marginal resource costs. In collective transport markets the situation is different: supply of collective transport requires some fixed costs (e.g. administration costs, storage facilities, non-vehicle network costs). Combination of fixed costs with constant marginal costs implies falling average costs (increasing returns to scale).

The model assumes that transport network costs are financed out of general government revenue. This reflects the insight from optimal tax theory[8] that there is in general no need to specifically make a link between tax revenues and government expenditures in one particular sector.[9] The standard version of TRENEN therefore contains no (balanced) budget requirement in any transport market or in any combination of transport markets. Instead, changes in tax revenues as caused by changes in transport policy are valued exogenously, and the exogenous value reflects an assumption on revenue use. For the purposes of our exercise, tax revenues get the same weight as consumer income. This is equivalent to assuming that individual lump sum redistribution of revenues is possible and that there are no distortions in the rest of the economy (see Proost and Van Dender (2001a) for more details, and see Section 4 for alternative assumptions). A budget constraint can be introduced into the model, as will be done in the AC pricing scenario.

3.2. The Implementation of the Pricing Scenarios in TRENEN

The first step in the construction of a policy scenario for AC pricing is to determine the revenue requirements, that is, the financial revenues to be raised through transport taxes.

The TRENEN-UIC dataset for the UK provides information on the cost side for collective transport modes (in particular, for urban metro, tram and bus and for non-urban bus and train), for 1995 and 2005. This information is sufficient to construct the revenue requirements for those modes (exclusive of road network costs, in the case of buses).

Information on the road network costs is harder to come by. Some reliable numbers come from Link and Suter (2001) who have studied the road network costs for Germany and Switzerland. However, the disaggregation of the road infrastructure costs to the case study level remains problematic. According to the analysis for Germany, the ratio of directly allocatable transport-specific revenues (i.e. exclusive of all VAT except VAT on fuel tax), to road infrastructure costs equals 1.59. Since no comparable data were available for the UK, we have used the German revenue/cost ratio for our simulations for the two UK regions.[10] The revenue component is computable from the TRENEN data for the reference equilibrium. We assume that the ratio (revenues over infrastructure costs in the road sector) is constant across the UK, in order to have a guesstimate for the road infrastructure costs.

Once the revenue requirements are determined, the following scenarios are computed.

- The reference equilibrium (REF)

This scenario uses the expected reference prices for 2005 in all transport markets. It serves as the benchmark to which the remaining scenarios are compared.

- Marginal social cost (MSC) pricing

This is the theoretical optimum obtained by the maximisation of the welfare function, allowing full differentiation of taxes across transport markets, without any budget constraint. From optimal tax theory we know that the resulting optimal prices will be equal to marginal social costs.

- Average cost (AC) pricing

Here the *modal budget* is financed by a uniform tax on all trips by that mode. In this exercise road network costs are allocated to all road user modes on the basis of their share in the total amount of vehicle-kilometres. Replacing the reference taxes by the taxes resulting from the modal revenue requirements will lead to demand changes. These will affect the revenue requirements. Model iterations therefore continue until revenue requirements, taxes and transport demand are mutually compatible[11] and a new equilibrium is reached.

3.3. Results

This section discusses the main issues concerning the effects of the pricing schemes for London[12] and the UK southeast region. First we discuss transport taxes, then traffic levels and composition, and finally welfare changes.

3.3.1. Tax Levels

Table 2 compares the reference taxes to total marginal external costs (TMEC). These consist of the sum of marginal external congestion costs (MECC) and other marginal external costs (pollution for small gasoline cars and diesel buses, accidents and noise), (MEPD). Marginal external congestion costs clearly are dominant in peak periods. The taxes include taxes on car ownership and on the use of cars and public transport. A negative tax corresponds with a subsidy. A negative tax is observed for bus transport in London. It means that the price paid by the bus users is lower than the marginal resource costs of producing the bus service. In London marginal external costs exceed taxes during peak hours. In off-peak periods current taxes cover more than the external costs in London, while for the southeast region this also is the case in peak hours.

Table 2. Marginal External Costs and Tax Levels in the Reference Situation – Partial Equilibrium Model (EURO/pkm, 2005).

	Tax	TMEC	MECC	MEPD	Tax	TMEC	MECC	MEPD
	Peak Car				Off-Peak Car			
London	0.118	0.503	0.447	0.056	0.108	0.090	0.035	0.055
UK southeast region	0.177	0.021	0.013	0.008	0.157	0.009	0.001	0.008
	Peak Bus				Off-Peak Bus			
London	−0.02	0.708	0.069	0.639	−0.010	0.398	0.005	0.393
UK southeast region	0	0.023	0.002	0.021	0.003	0.055	0.001	0.054

Note: Negative taxes mean subsidies.

Table 3. Tax Levels for Various Pricing Scenarios – Partial Equilibrium Model (EURO/pkm, 2005).

	REF	MSC	AC	REF	MSC	AC
	Peak Car			Off-Peak Car		
London	0.118	1.000	0.105	0.108	0.840	0.105
UK southeast region	0.177	0.114	0.060	0.157	0.100	0.060
	Peak Bus			Off-Peak Bus		
London	−0.02	0.785	0.234	−0.010	0.517	0.231
UK southeast region	0	0.043	0.082	0.003	0.073	0.254

Table 3 is an overview of the tax levels in car[13] and public transport markets, for the two pricing scenarios. For car markets, we take the example of a small petrol car with one driver-occupant. For public transport markets, we look at bus markets.

In London the optimal taxes under MSC pricing are higher than the reference taxes for peak and off-peak car trips. This is not the case in southeast England, which has low average congestion levels in the reference equilibrium. This result may be explained by the geographical scale of the case study for the southeast region. Taking large networks into account may tend to spread out congestion levels.

It is clear from Table 3 that combining cost recovery by mode with AC pricing leads to substantial car tax reductions and (very) large bus tax increases in all cases. The resulting impact on the modal split is described in the next subsection.

Table 4. Traffic Level Index (PCU) and Transport Demand Index (pkm) under Various Pricing Scenarios – Partial Equilibrium Model (2005, REF = 1).

	REF	MSC	AC
Traffic level index (PCU)(REF = 1)			
London	1	0.76	1.06
UK southeast region	1	1.01	1.11
Transport demand index (pkm)(REF = 1)			
London	1	0.88	1.03
UK southeast region	1	1.00	0.96

3.3.2. Traffic Level and Composition

Table 4 shows the impact of the pricing mechanisms on traffic levels passenger car units (PCU) and on transport demand (pkm). In London MSC pricing reduces travel demand (pkm) and traffic flows (PCU) in comparison to the reference situation. In the southeast region traffic flows and transport demand are almost unaffected. MSC pricing leads to revenues in excess of the revenue requirements specified for AC pricing. Under MSC pricing there is no longer a justification for subsidising public transport beyond the level of fixed costs.[14] The efficient modal split, i.e. the modal split that maximises social welfare, is obtained by pricing all modes at their marginal social cost.

AC pricing leads to an *increase* in traffic levels, since the change in taxation favours a shift from public transport to car transport and taxes are on average reduced with respect to the reference equilibrium. In London the increase in PCU is larger than for pkm, because of the modal shift towards car trips, away from collective modes. This is the consequence of the relatively high revenue requirement (hence, relatively high taxes) for collective modes, and the relatively low revenue requirement in car markets. This feature of *modal* AC pricing shows that defining budget requirements in narrow sets of transport markets may have undesirable effects on the modal split. The simple AC pricing scheme performs badly both in terms of aggregate travel demand and in terms of the modal split for a given level of demand. In the southeast region transport demand is reduced, but the shift towards car modes leads to a larger traffic level than in the reference situation.

Table 5 shows the share of the traffic volume (PCU) taking place during peak hours. As can be seen, this share is less sensitive to the pricing scheme

Table 5. Share of Peak Period PCU – Partial Equilibrium Model (2005).

	REF (%)	MSC (%)	AC (%)
London	68.1	69.1	66.9
UK southeast region	69.5	69.3	69.8

than is the total traffic volume. A large share of peak transport consists of commuting transport, which is characterised by lower prices elasticities.

In the urban case study, the impact of the two pricing schemes on freight transport is small, and the directions of change are similar to those of passenger car transport. In the southeast region, AC pricing decreases the modal share of rail freight, in comparison with the reference situation.

3.4. Welfare Impacts

Table 6 shows the welfare changes induced by the pricing scenarios for the various cases. In the absence of distortions in the rest of the economy, welfare is given by the sum of the utility obtained by households from the consumption of private goods and the utility of freight transport users, minus the external environmental and accident costs.

Household utility is a function of three elements: the prices of non-transport goods (which are assumed to be constant), the generalised costs of all passenger transport modes and household income. Household income takes into account recycled tax revenue.

In line with the theoretical analysis, MSC pricing outperforms AC pricing. The introduction of a budget constraint reduces the efficiency effects of transport pricing systems. Second, the way in which this constraint is met, has further consequences for the welfare effects.[15] The welfare effects are measured relative to the generalised income (potential gross labour income) that is about 1.5 times GDP.

Interestingly, AC pricing leads to a reduction of welfare with respect to the reference situation in both cases. While the size of the reduction varies substantially between cases, the basic reasons for the welfare reductions are the same. They are twofold.

First, the current transport prices go some way in the direction of a second-best pricing structure. Under-priced passenger car transport (from the social point of view) is often combined with subsidised public transport, so that relative price distortions are reduced. Such a policy is not feasible

Table 6. Welfare Impacts of Pricing Scenarios – Partial Equilibrium Model (2005, % change with respect to REF).

	MSC	AC
London	+2.70	−0.76
UK southeast region	+0.55	−1.89

under the *modal* budget requirements used in the AC pricing simulations. Taxes for each mode are only determined by the modal revenue requirement, so that no account can be taken of prices in substitute modes.

Second, the modal budget constraints require less revenue than is raised in the reference situation. This means that the revenues from current transport taxes are higher than what is required in the transport sector. Optimal commodity tax theory shows that, if transport demand is relatively inelastic, revenue raising in that sector tends to limit the efficiency cost of collecting the required total amount of government revenue.[16] The fact that the transport sector at present is 'revenue positive' may then be justified from the optimal taxation point of view, although there is no guarantee that relative prices or the size of the surplus are anywhere near optimal.

If the revenue requirement were increased above the transport-related requirement, AC pricing could, but need not, perform better than the reference price structure. As the peak-period taxes from AC pricing approach the peak period external costs, the performance of AC improves. This improvement will be counteracted to some extent by the growing deviation between off-peak taxes and off-peak external costs. However, since peak-period congestion costs are the dominant externality, a net improvement of welfare should be expected.

The variation in results between cases depends on the degree of cost coverage of collective modes in the reference situation, and on the degree to which the new budget constraint allows sufficient differentiation of prices with respect to transport externalities.

4. THE INDIRECT EFFECTS OF ALTERNATIVE PRICING APPROACHES

The aim of this section is to extend the analysis of Section 3 in several ways. We claim that a full evaluation of the efficiency and equity impacts of transport policies requires a general equilibrium rather than a partial equilibrium approach. There are several reasons for this. First of all, the general equilibrium approach allows us to take into account the impacts on all sectors of the economy, and not only on the transport sectors. Second, it offers the advantage of considering transport taxes within the framework of global tax policy. Account can be taken of the interactions between transport policies and public finance. While Section 3 assumes that there are no pre-existing distortionary taxes in the rest of the economy, this assumption is dropped here. The ultimate welfare impact of transport policies depends on how the budget deficits of the transport sector are financed by taxes on other sectors or how the excess tax revenues generated in the transport sector are used to decrease taxes in the rest of the economy. This is important both for the efficiency and the equity effects of transport policies. In Section 3 only efficiency considerations were taken into account. Here we also consider the equity impacts.

In this section we present results obtained with a computable general equilibrium (CGE) model for Belgium. Section 4.1 provides a short model description. The policy scenarios are described in Section 4.2. Section 4.3 makes a comparison between MSC pricing and "pure" AC pricing, and between two ways of using tax revenues generated in the transport sector: via the labour income tax and via the social security transfers. The focus of the presentation lies on the distributional impacts of the policy scenarios.

4.1. The CGE Model for Belgium

The CGE model for Belgium is a static model for a small open economy, with a medium term time horizon. It considers four types of economic agents: five consumer groups, fourteen main production sectors, the government and the foreign sector. The model is an extension of Mayeres (2000)[17] in that it includes five consumer groups, corresponding with the quintiles of the Belgian household budget survey, instead of one representative consumer group. Two persons belonging to a different consumer group differ in terms of their productivity, their tastes and their share in the total endowment of capital goods and the government transfers. Persons belonging to the same consumer group are however assumed to be identical in terms of their needs. The second difference with respect to Mayeres (2000) concerns the production technology of the public transport sectors: the model takes into account the existence of fixed costs, rather than assuming constant returns to scale technology. For the other production sectors we assume a constant returns to scale technology, with freight transport as one of the inputs in the production process, together with labour, capital, energy and other commodities.

The CGE model includes several transport commodities, as presented in Table 7. A distinction is made between passenger and freight transport, between various transport modes, between vehicle types and for some transport modes between peak and off-peak transport.

The model considers four types of externalities: congestion,[18] air pollution (including global warming), accidents and road damage externalities. Air pollution and accidents are assumed not to affect the behaviour of the economic agents,[19] but have a separable impact on welfare. The modelling of congestion accounts not only for its negative impact on the consumers' welfare, but also for its influence on their transport choices. This also implies that the value of a marginal time saving is determined endogenously in the model. Congestion also reduces the productivity of transport labour in the production sectors. The inclusion of the road damage externalities is a third extension with respect to Mayeres (2000).

Table 8 presents the marginal external costs[20] of the various transport modes in the benchmark equilibrium, which corresponds with the situation in Belgium in 1990.[21] It also compares the marginal external costs with the taxes paid in order to give an idea of the distortions in the benchmark. For

Table 7. Transport in the CGE Model for Belgium.

Passenger Transport[a]			Freight Transport[b]		
	Private	Business		Domestic	Export or Import Related, Transit
Car			Road		
Gasoline	X	X	Gasoline van	X	
Diesel	X	X	Diesel van	X	
LPG	X	X	Truck	X	X
Bus, tram, metro	X		Rail	X	X
Rail	X	X	Inland navigation	X	X
Non-motorised	X				

[a]For all passenger transport modes a distinction is made between peak and off-peak transport.
[b]The split between peak and off-peak transport is made only for road transport.

Table 8. The Marginal External Costs and Taxes in the Benchmark Equilibrium – CGE Model for Belgium (1990).

		Marginal External Cost (EURO /vkm)	Share in Marginal External Costs				Tax[a] (EURO/ vkm)
			Road damage (%)	Congestion (%)	Air pollution (%)	Accidents (%)	
Passenger transport							
Peak	Gasoline car	0.28	0	83	6	10	0.10 [0.04]
	Diesel car	0.33	0	73	18	9	0.06 [0.02]
	Tram	0.53	11	89	0	0	−1.77
	Bus	1.10	6	43	41	11	−1.77
	Rail (diesel)	0.19	0	0	100	0	−8.99 [−9.66]
Off-peak	Gasoline car	0.09	0	49	20	31	0.10 [0.04]
	Diesel car	0.13	0	34	44	22	0.06 [0.02]
	Tram	0.15	40	60	0	0	−1.41
	Bus	0.72	9	13	62	16	−1.41
	Rail (diesel)	0.19	0	0	100	0	−1.61 [−1.73]
Freight transport							
Truck – peak		0.86	7	55	36	2	0.13
Truck – off-peak		0.48	13	19	65	3	0.13
Inland navigation		0.01	0	0	100	0	n.a.
Rail (diesel)		0.52	0	0	100	0	−0.84

[a]The figure in brackets refer to transport for business purposes.

peak road transport congestion accounts for the largest share in the external costs. In the off-peak period air pollution is the most important external cost category for diesel vehicles, while accident costs form the largest category for gasoline vehicles. For all transport modes there is a large divergence between the tax and the marginal external costs, which implies that MSC pricing will lead to substantial price changes. In the case of public transport the subsidies[22] related to the provision of the transport services are high, resulting in a negative tax.

Before turning to the simulation results, attention should be drawn to a number of limitations of the CGE model. First, the model cannot be used to simulate the location decisions of the households and firms. Therefore, our analysis does not capture the equity impacts of a change in land use. Second, only a limited number of consumer groups is considered. A further disaggregation of the consumer groups would generate additional insights in the equity impacts of the policy reforms. Third, the model does not make a distinction between different trip purposes. Since different trip purposes have a different relationship with labour supply, this will affect optimal taxation[23] (see Parry and Bento, 2001; Calthrop, 2001; Van Dender, 2003). These issues are not considered here. Fourth, the shift towards cleaner vehicle types[24] or vehicles with lower road damage costs is not modelled. This implies that not all effects of the pricing reforms on the externalities are included.

4.2. The Policy Scenarios

The CGE model for Belgium is used to calculate the welfare effects of MSC pricing and to compare them with the effects of AC pricing. Since these policies have an impact on the government budget, their full welfare impact can be assessed only if one considers the accompanying measures by which the government achieves budget neutrality. This is necessary for evaluating both the efficiency and the equity impacts. Here we consider two such instruments, namely the labour income tax and the social security transfers.[25] Table 9 summarises the four scenarios for which the welfare effects are calculated.

4.2.1. Marginal Social Cost Pricing

Scenarios 1 and 2 concern MSC pricing for all transport sectors except inland navigation. All existing taxes (including VAT) and the subsidies related to the variable operating costs of public transport are abolished. An

Table 9. Overview of the Policy Scenarios – CGE Model for Belgium.

		Budget Neutrality Ensured by	
		Labour Income Tax	Social Security Transfers
Pricing rule	MSC	Scenario 1	Scenario 2
	AC	Scenario 3	Scenario 4

externality tax is introduced ensuring that each transport user pays his marginal social cost. Both domestic and transit transport are subject to the tax reform.

4.2.2. Average Cost Pricing

Scenarios 3 and 4 introduce AC pricing for three transport sectors: road, rail and other public transport. Due to the lack of reliable data for inland navigation, this sector is not included in the exercise. AC pricing is defined as balancing the financial budget for each of the three transport sectors. For road transport the financial costs equal the infrastructure costs (excl. taxes). For public transport they equal the sum of infrastructure and supplier operating costs (excl. taxes). The financial costs and revenues are calculated as much as possible according to the methodology of Link et al. (2000, 2002).

In the AC scenarios all existing transport taxes (except the VAT) and subsidies are set equal to zero. The VAT rate is set at the standard rate.[26] In the case of road transport AC pricing is introduced by means of an undifferentiated tax per vkm for car, truck and bus. No distinction is made between heavy and light road vehicles. Transit transport is also subject to the tax reform. For rail transport an undifferentiated tax is imposed on pkm and tkm. For the other public transport modes an undifferentiated tax per pkm is used.

In all four scenarios budget neutrality is imposed. We consider two alternative instruments to achieve this. In Scenarios 1 and 3 budget neutrality is obtained by means of the labour income tax. In Scenarios 2 and 4 the social security transfers are changed. In all cases an equal percentage change in the revenue-recycling instrument is assumed for all quintiles. It is evident that this assumption affects the distributional impacts of the scenarios and that different assumptions may lead to different conclusions.

4.3. Results

4.3.1. Transport Prices, Transport Demand and Marginal External Costs

Tables 10–12 summarise the impact of MSC and AC pricing on prices, demand and marginal external costs. The tables only present the results for scenarios 1 and 3 in which the labour income tax is used as the revenue recycling instrument. The impacts are similar when the social security transfers are changed instead.

Marginal Social Cost Pricing. With MSC pricing all existing transport taxes and the subsidies related to variable operating costs are abolished, and replaced by a tax per vkm that reflects the marginal external costs. It should be noted that the CGE model is a model for a second-best economy in which the government needs to use distortionary taxes to finance its budget. Therefore, the MSC pricing that is considered here is in general not the optimal[27] pricing policy. Note also that all other taxes (except the revenue recycling instruments) are assumed to remain constant.

Road transport MSC pricing entails a substantial increase in the tax per vkm in the peak period, reflecting the high congestion costs. In the off-peak period the tax is also raised for most road transport modes, but less so. The tax is differentiated according to vehicle type. For example, diesel cars are now subject to a higher tax than gasoline cars, since they are associated with higher air pollution costs. The MSC scenarios also imply a tax increase for the input of vkm by the public transport companies.

Table 10 presents the resulting effect on transport prices. In most cases transport prices increase. The increases are higher in the peak than in the off-peak. The increase in the price of public transport reflects not only the internalisation of the external costs, but also the abolition of the variable subsidies. Total transport demand falls both for passenger (−8.2%) and freight transport (−9.5%). The reduction in transport demand means that people respond to this type of pricing by reconsidering their transport decisions, for example, by abolishing some trips, by choosing destinations that are closer by, or by rethinking the organisation of production. As is shown in Table 11, MSC pricing also leads to a shift from peak to off-peak transport both for passenger and freight transport, and from private to public transport (except for off-peak passenger transport).

The impact on the marginal external costs is summarised in Table 12. MSC pricing raises average road speed during the peak, which is the main explanation of the fall in the marginal external costs in this period (−30%

Table 10. The Effect of the Policy Reforms on Transport Prices – CGE Model for Belgium (1990).

Belgium	Benchmark	Scenario 3	Scenario 1
		MSC + Lower Labour Income Tax	AC + Higher Labour Income Tax
Price Passenger Transport	(EURO/pkm)	Percentage Change w.r.t. benchmark	
Peak			
Gasoline car – committed[a]	0.29	21	−16
Gasoline car – suppl.[a]	0.13	84	−19
Diesel car – committed	0.19	69	−12
Diesel car – suppl.	0.08	209	−8
Bus, tram, metro	0.06	89	154
Rail	0.06	73	764
Off-peak			
Gasoline car – committed	0.29	−8	−16
Gasoline car – suppl.	0.13	16	−19
Diesel car – committed	0.19	23	−12
Diesel car – suppl.	0.08	98	−8
Bus, tram, metro	0.06	146	193
Rail	0.06	72	769
Price Freight Transport	(EURO/tkm)	Percentage Change w.r.t. benchmark	
Truck			
Peak – committed	0.17	40	−10
Peak – suppl.	0.17	111	−9
Off-peak – committed	0.16	27	−11
Off-peak – suppl.	0.16	89	−12
Rail	0.05	7	349

[a]The distinction between committed and supplementary mileage allows us to model the link between car ownership and car use. The CGE model assumes that owning a car implies a certain minimum mileage. This is reflected in the committed mileage, which is proportional to the vehicle stock. The costs of committed mileage include the ownership and running costs per km. The consumers can choose to drive more than the minimum mileage per car. This is captured by the supplementary mileage, whose cost includes only running costs.

Table 11. The Effect of the Policy Reforms on Transport Demand – CGE Model for Belgium (1990).

Belgium	Benchmark	Scenario 1	Scenario 3
		MSC + Lower Labour Income Tax	AC + Higher Labour Income Tax
Passenger transport	mio pkm/year	Percentage Change w.r.t. benchmark	
Peak	36,532	−12.89	−2.38
car	29,308	−14.28	4.68
bus, tram, metro	4,239	−3.98	−18.82
rail	2,985	−11.93	−48.33
Off-peak	59,684	−5.42	+2.73
car	51,813	−3.36	8.75
bus, tram, metro	4,317	−20.38	−26.19
rail	3,554	−17.30	−50.04
Freight transport	mio tkm/year	Percentage Change w.r.t. benchmark	
Road – peak	7,485	−15.04	10.34
Road – off-peak	32,715	−11.12	11.07
Rail	8,354	6.51	−81.28

for the passenger modes, and −20% for trucks). The impact in the off-peak period is much smaller.

Average Cost Pricing. In the initial equilibrium the financial cost coverage rate, defined as the ratio of revenue over financial costs, equals 2.54 for road transport, 0.28 for rail and 0.37 for other public transport. The two AC pricing scenarios therefore imply a reduction in the taxes on road transport and a substantial increase in the taxes on public transport. As a result, the money price of car and truck transport falls, while that of public transport increases considerably (Table 10). Table 11 gives the resulting impacts on transport demand.[28] For passenger transport there is a shift from the peak to the off-peak and from public transport to private transport. The share of public transport becomes much smaller. Similar impacts are observed for freight transport. Total demand rises by approximately 0.8% for passenger transport and falls by approximately 3.1% for freight transport.

Table 12 shows how the AC policies affect the marginal external costs of transport. Given the shift towards road transport, average road speed falls both in the peak and the off-peak period, which leads to an increase in the marginal external costs of road transport, especially during the peak.

Table 12. The Effect of the Policy Reforms on the Marginal External Costs of Transport – CGE Model for Belgium (1990).

Belgium	Benchmark	Scenario 1	Scenario 3
		MSC + Lower Labour Income Tax	AC + Higher Labour Income Tax
Marginal external cost (EURO/vkm)			
Peak			
Gasoline car	0.28	0.19	0.32
Diesel car	0.33	0.23	0.36
Tram	0.53	0.35	0.60
Bus	1.10	0.92	1.16
Truck	0.86	0.68	0.93
Off-peak			
Gasoline car	0.09	0.09	0.10
Diesel car	0.13	0.13	0.14
Tram	0.15	0.14	0.17
Bus	0.72	0.71	0.73
Truck	0.48	0.47	0.50
Rail			
Passenger – el.	0.00	0.00	0.00
Passenger – diesel	0.19	0.19	0.19
Freight – el.	0.00	0.00	0.00
Freight – diesel	0.52	0.52	0.51

The Transport Accounts. Tables 13–15 present the impact of the policy scenarios on the transport accounts. The transport accounts give an overview of the costs and revenues for each mode of transport. They are constructed according to the methodology developed by Link et al. (2000, 2002). On the cost side a distinction is made between infrastructure costs, supplier operating costs and external accident and environmental costs. The accounts also present information on the user and internal accident costs. The user costs refer to the delay costs caused by and incurred by transport users. The revenue side includes the net revenues from transport pricing, charging and taxation.

The impact of the MSC and AC pricing scenarios on the accounts is computed by means of the CGE model for Belgium.[29] Since the transport accounts are almost not affected by the choice of the revenue recycling instruments considered here, the tables only present the accounts for MSC and AC pricing when they are accompanied by a change in the labour income tax.

Table 13. Road Transport Account – Belgium (1990).

	Benchmark	Scenario 1	Scenario 3
		MSC + Lower Labour Income Tax	AC + Higher Labour Income Tax
Costs (mio €)			
Infrastructure costs (excl. taxes)	1,797	1,812	1,815
Capital costs	1,198	1,241	1,196
Running costs	599	571	619
Fixed	300	300	300
Variable	299	271	319
External accident costs	2,198	2,057	2,341
Environmental costs (change w.r.t. benchmark)		−537	255
Costs (additional information)			
User costs (time)	1,531	758	1,770
Internal accident costs	5,006	4,683	5,331
Revenues (mio €)	4,562	10,382	1,833
Kilometre charge	0	11,912	1,833
Circulation tax	868		
Fuel duty	2,535		
VAT	1,158	−1,530	
Revenues/Financial Costs	2.54	5.73	1.01

MSC pricing raises the financial cost coverage rate for the three transport sectors. For road and public transport other than rail it more than doubles. On the cost side, the change in financial, accident and environmental costs is a consequence of the change in transport demand, as described in Table 11.

Note that the financial cost coverage rate for the MSC scenarios can to some extent be used as guidance for determining the optimal surplus/deficit by mode. However, this information is not sufficient for determining optimal policies, since there is no guarantee that the associated tax structure is optimal.

The AC pricing scenarios ensure a financial cost coverage rate of 1 for the three transport sectors. For the road sector, the uniform levy leads to an

Table 14. Public Transport (other than rail) Account – Belgium (1990).

	Benchmark	Scenario 1	Scenario 3
		MSC + Lower Labour Income Tax	AC + Higher Labour Income Tax
Costs (mio €)			
Total financial costs	1,149	966	951
Infrastructure costs (excl. taxes)	30	31	30
Supplier operating costs (excl. taxes)	1,118	934	921
Vehicle related	741	581	589
Other	378	353	332
External accident costs	32	27	25
Environmental costs (change w.r.t. benchmark)		−19	−29
Costs (additional information)			
User costs (time)	226	110	216
Internal accident costs	61	52	47
Revenues (mio €)	430	799	948
Tax on pkm or tkm	0	0	191
Excises paid by operators	4	4	4
Tariff revenue	479	958	754
Taxes on tariffs	−53	−163	0
Revenues (additional information)			
Subsidies related to variable operating costs	506	0	0
Revenues/Financial Costs	0.37	0.83	1.00

increase in revenues. There is also a small increase in infrastructure costs, due to higher demand for road transport. The higher transport demand, accompanied by a shift to private transport, also increases the accident and environmental costs of the road sector. For public transport, the higher

Table 15. Rail Account – Belgium (1990).

	Benchmark	Scenario 1	Scenario 3
		MSC + Lower Labour Income Tax	AC + Higher Labour Income Tax
Costs (mio €)			
Total financial costs	2,608	2,486	1,923
Infrastructure costs (excl. taxes)	872	891	871
Supplier operating costs (excl. taxes)	1,736	1,595	1,052
Vehicle related	700	634	276
Other	1,035	962	776
External accident costs	0	0	0
Environmental costs (change w.r.t. benchmark)		0	−8
Costs (additional information)			
User costs (time)	0	0	0
Internal accident costs	0	0	0
Revenues (mio €)	723	925	1,927
Excises paid by operator	1	1	1
Tariff revenue	759	1,012	424
Taxes on tariffs	−37	−89	1,502
Revenues (additional information)			
Subsidies related to variable operating costs	350	0	0
Revenues/Financial Costs	0.28	0.37	1.00

prices lead to more revenue, but also to lower costs due to the reduction in demand for public transport.

The transport accounts summarise some of the effects of the policy scenarios. However, they give no indication of whether AC pricing and/or MSC pricing constitute an improvement with respect to the benchmark equilibrium. Nor do they allow to make a choice between revenue recycling instruments. This is because they do not contain all elements necessary for a social cost benefit analysis. The transport accounts do not take into account the benefits that people derive from transport or their user costs, nor do they allow to assess the effects on the different income groups. Moreover, they do not take into account how a higher deficit in the transport sector is financed or how additional revenue is used. Therefore, additional analysis is required. Here we use the CGE model to calculate the social welfare impacts of the policy reforms.

4.3.2. The Impact on Welfare

Table 16 summarises the welfare effects of the policy reforms for the household income quintiles. Quintile 1 represents the poorest households, while quintile 5 represents the richest households. The welfare impact on the quintiles is measured by means of the equivalent gain: the increase in the initial equivalent income of an individual which is equivalent to implementing the policy reform. In the table it is presented as the percentage increase in the initial equivalent income of the individual. The effect on social welfare is described by the social equivalent gain. This is defined as the change in each individual's original equivalent income that would produce a level of social welfare equal to that obtained in the post-reform equilibrium. The social desirability of a policy depends not only on its efficiency, but also on its equity impact. Hence we present the social equivalent gain for two degrees of inequality aversion. In the first case only efficiency matters. We also present the social welfare change for a medium degree of inequality aversion. In this case the marginal social welfare weight of people belonging to the richest quintile is approximately 70% of those belonging to the poorest quintile.

Marginal Social Cost Pricing. Since the MSC scenarios raise the revenue collected by the government, the full welfare assessment needs to take into account how this revenue is used. In Scenario 1 the labour income tax is reduced by 10% for all quintiles. In Scenario 2 the extra revenue is used to increase the social security transfers by 11%.

In both MSC scenarios the impact on social welfare is positive. The reason for this is similar as in Section 3. However, not all quintiles benefit to the same extent from the policy reforms. Moreover, the welfare impacts on the quintiles are quite different in the two MSC scenarios. When the extra revenue is returned through higher transfers (Scenario 2), the welfare gains become lower as the income of the quintiles rises. The poorer quintiles benefit most from the higher transfers, since they make up a higher

Table 16. The Welfare Effects of the Policy Reforms – CGE Model for Belgium.

	Benchmark	Scenario 1	Scenario 2	Scenario 3	Scenario 4
		MSC + Lower Labour Income Tax	MSC + Higher Social Security Transfers	AC + Higher Labour Income Tax	AC + Lower Social Security Transfers
Equivalent income (EURO/person/year)			Percentage Change w.r.t. benchmark		
Quintile 1	18,586	0.47	3.88	−0.78	−0.97
Quintile 2	22,260	0.03	2.21	−0.04	−0.16
Quintile 3	25,027	−0.16	0.75	−0.24	−0.29
Quintile 4	28,330	0.22	0.00	−0.20	−0.19
Quintile 5	35,579	1.45	−0.51	−0.49	−0.38
Social equivalent gain			(EURO/person/year)		
Efficiency		160.66	148.89	−92.71	−92.08
Medium degree of inequality aversion		142.50	179.17	−89.56	−91.74

share of their income. In this scenario the two richest quintiles do not benefit from the policy reform: they pay higher transport taxes, but benefit only to a small extent from the redistribution of the extra government revenues.

In Scenario 1 all quintiles gain, except quintile 3. This quintile consumes a lot of car transport and does not benefit as much as the richer quintiles from the reduction in the labour income tax. The highest welfare gain is observed for quintile 5. While consuming relatively a lot of transport, this quintile benefits most from the lower labour income tax and from the reduction in the externalities.

While the transport account is similar in Scenarios 1 and 2, the impact on social welfare is not. It depends on the revenue-recycling instrument that is used, and on the inequality aversion of society. When only efficiency considerations are important, the labour income tax is preferred as revenue recycling instrument. When a higher weight is given to the poorer quintiles it is better to recycle the revenue through higher transfers.

Average Cost Pricing. AC pricing leads to a reduction in government revenue. In Scenario 3 this is financed by an increase in the labour income tax by 0.5% for all quintiles. In Scenario 4, the social security transfers are reduced by 1% for all quintiles.

Both AC scenarios reduce welfare for all quintiles. Consequently they both lead to a social welfare loss. This shows clearly that balancing the financial part of the transport accounts is not an objective that one should aim at. This is in line with the conclusions of Section 3. Table 16 also shows that AC pricing cannot be defended because of equity reasons, since all income groups become worse off. These findings are a confirmation of the theoretical discussion in Mayeres et al. (2001).

Within each AC scenario the differential impact on the quintiles can be explained, inter alia, by their share in the consumption of the transport goods, their share in the total social security transfers or labour income, the level of initial taxation and the quintiles' valuation of the reduction in the externalities. For example, the relatively high welfare loss of quintile 1 in Scenario 3 is due to its large share in the consumption of public transport which becomes considerably more expensive. The high welfare loss of quintile 5 is due to the high share of labour income in its income and the high labour income tax of this quintile in the benchmark equilibrium.

The difference in welfare impact between Scenarios 3 and 4 is due to choice of budget-neutralising instrument. When the social security transfers are reduced, the welfare losses for quintiles 1–3 are higher than when the labour income tax is increased. This is because the social security transfer accounts for a larger share of their income. The share of labour income is relatively smaller for these quintiles, as is the labour income tax rate.

Social welfare is reduced in both AC scenarios, as is reflected in the negative social equivalent gain. The social equivalent loss does not differ a lot between the two revenue-recycling strategies. This is because the required changes in the labour income tax and the social security transfers are relatively small. With AC pricing the impact on welfare is dominated by the change in the transport taxes.

5. CONCLUSIONS

Our case studies indicate that AC pricing leads to higher prices for public transport and lower prices for car transport. This results in an increase in traffic levels and a modal shift towards car transport, away from public transport. With MSC pricing there is a shift from the peak to the off-peak period and from cars and trucks to the collective modes.

Both the partial and the general equilibrium analysis show that the requirement of modal budget balance reduces welfare when this requirement is met through AC pricing. The welfare cost of AC pricing relative to MSC pricing may be substantial. The general equilibrium analysis for Belgium also shows that AC pricing based on financial costs leads to welfare losses for all income groups considered in the study with the biggest percentage loss for the poorest group. This indicates that AC pricing cannot be justified on equity grounds.

Of course, there are other ways of defining AC pricing schemes than the ones analysed here, and alternative definitions may produce better results. These alternative schemes will however become more complex, and they will still perform worse than marginal-cost-based pricing approaches. From related economic literature we know that when the budget is met through Ramsey pricing, rather than AC pricing, this is most often welfare improving (see, for example, Proost and Van Dender, 2004), though the imposition of a budget constraint still involves an important welfare cost in comparison with pricing schemes that do not impose such a constraint. Another option may be to use two-part tariffs (see, for example, De Borger, 2001). The welfare effect therefore depends not only on the presence of a budget constraint but also on the flexibility with which that constraint can be met.

MSC pricing generally increases social welfare. The CGE model for Belgium shows that in general not all groups are affected equally by MSC pricing. The equity impacts depend on how budget neutrality is ensured.[30] The Belgian CGE model, which considers several income groups, shows that when society becomes more inequality averse, the revenue recycling instrument that is more beneficial to the poorer income groups will be preferred. One can conclude that the revenue recycling instruments have an important role to play in enhancing the political acceptability of transport pricing. Our results show that it is possible to use a pricing scheme based on MSC and to complement it by a carefully designed revenue recycling scheme such that the majority of people gain from the tax reform.

It should be noted that the relative welfare effects of MSC and AC pricing depend on the transport situation in the country or region under study. In countries where congestion is the dominant marginal external cost of transport, transport instruments that tackle this problem efficiently have an advantage over the others. Therefore, instruments which do not make a distinction between congested and uncongested situations get a large penalty. A second determining factor is the ratio of transport revenue to financial costs in the reference equilibrium. A different ratio means that the alternative pricing instruments will have a different implication for the transport accounts and government budget.[31]

NOTES

1. The research reported in this chapter was funded by the UNITE project of the 5th Framework RTD Programme of the European Commission and by a postdoctoral grant of the Fund for Scientific Research – Flanders. The authors wish to thank Bryan Matthews and Chris Nash for their valuable comments and suggestions. The authors remain fully responsible for the contents of the chapter. The conclusions presented here do not necessarily reflect the official views of the Belgian Federal Planning Bureau.

2. For example, in the case of car use, the marginal resource costs include the fuel, vehicle, maintenance, insurance, parking costs, etc. exclusive of taxes. The marginal resource costs of public transport use equal marginal supplier operating costs before subsidies.

3. When these conditions do not hold, additional elements should be included in the cost–benefit analysis. For example, when the price is higher than the short run MSC, the social benefit of capacity expansion is increased: a capacity expansion would help to bring the volume of transport to a higher and more optimal level.

4. Ramsey pricing and two-part tariffs are examples of such pricing rules.

5. We could add external costs to the total costs to be covered by each mode in the average cost scenario. This could improve slightly the efficiency results of the average cost scenario as regards those external costs for which marginal value is close to the average value (for example, air pollution).

6. Since the marginal external cost studies described in the rest of the book were not yet available at the time of the study, the calculation of the marginal external costs is based on the methodology of Mayeres and Van Dender (2001). We expect that the general results continue to hold when the more recent marginal external cost data are included in the analysis.

7. In line with the medium run horizon of the model, car ownership costs are expressed on a per vehicle-kilometre basis.

8. See, for example, Mayeres and Proost (1997).

9. Making such a link (hypothecation or earmarking) leads to constraints on the optimal tax configuration. As long as information is costless, suppliers are cost-minimisers and politicians and agencies maximise the welfare of the citizens, such constraints are at best harmless, but will in general entail efficiency costs.

10. Sensitivity analysis shows that the results do not strongly depend on these ratios.

11. This need for an iterative procedure shows that calculating consistent average cost scenarios is as complex, in terms of model requirements, as computing Ramsey prices. This casts some doubt on assertions that average cost pricing measures are easier to implement because they are less complex.

12. The results for London may be typical for large metropolitan areas, but cannot be taken as representative for all urban areas.

13. Car taxes are different in the peak and off-peak period even though the tax system does not distinguish between time of day. The difference is explained by the fact that the TRENEN model takes into account the difference in fuel consumption between peak and off-peak.

14. There is one exception: economics of density in public transport (Mohring effect). This implies that every extra passenger allows to improve frequency of service. Therefore, there is a positive externality in public transport. The marginal social cost thus equals the sum of the marginal operating costs and the marginal external costs (congestion, air pollution, accidents) minus the external benefit of a more frequent service (see also Chapter 5 on user costs).

15. Proost and Van Dender (2004) show that Ramsey social cost pricing performs better than AC pricing, because it allows for a differentiation of transport prices across modes and times of day. In that case the effect of a budget constraint at the level of the transport sector as a whole is mitigated to a considerable degree. Nevertheless, the welfare cost of imposing a budget constraint remains substantial in comparison with an MSC pricing scheme.

16. We neglect here potential interactions with other distorted markets (labour market).

17. For a detailed discussion of the CGE model the reader is referred to Mayeres (1999, 2000).

18. The model only considers the time costs of congestion. The effects of congestion on the emission factors or the accident risks is not yet incorporated.

19. In reality they also affect the consumption and production choices. Such feedback effects are not yet included in the model. The feedback effect of the

health-related benefits of environmental policies is studied by Mayeres and Van Regemorter (2003).

20. In order to simplify the presentation, we have included the marginal road damage costs in the external costs. It should be noted however, that these costs are not external costs but costs to society that are not or only partly reflected by the existing pricing schemes of publicly owned roads.

21. There are two exceptions, however. First, the emission characteristics of cars and trucks are those observed in 2000. Secondly, the data by income class are based on the Belgian expenditure survey for 1995. For details on the data set and the model calibration, the reader is referred to Mayeres (1999) and Mayeres and Proost (2002).

22. Table 8 only includes the public transport subsidies related to variable operating costs.

23. In general a distinction can be made between commuting and non-commuting trips. If these two trip types can be taxed differently, theory suggests that one should tax commuting trips, which are complementary with labour, less than non-commuting trips, which are complementary with leisure. Imposing uniform taxes on the two trip purposes may entail substantial welfare losses under certain conditions (see, for example, Van Dender, 2003).

24. The model only considers the choice between diesel and gasoline cars, for a given emission technology and fuel efficiency.

25. Various alternatives, such as infrastructure investment, a change in the supply or quality of public transport, etc., could be considered. The efficiency and equity effects will in general depend on the instrument that is chosen.

26. The treatment of VAT in these scenarios is in line with the methodology of Link et al. (2000) for defining revenues in transport accounts.

27. For the optimal tax and investment rules in a second-best economy in the presence of externalities, see Mayeres and Proost (1997).

28. The average cost scenarios are quite extreme scenarios. The resulting demand changes should be regarded with caution. They should be considered as very rough estimates for the cases with large price changes.

29. For the environmental costs Tables 13–15 present the change with respect to the reference equilibrium, rather than the total environmental costs. For the valuation of emissions we have information only about the marginal willingness-to-pay for emission reductions. These values can be used only to evaluate relatively small changes in emissions. It would therefore be incorrect to use them to calculate the total environmental costs.

30. This would also hold when equity is considered in terms of other dimensions than income, e.g. when a distinction is made according to urban and non-urban households as in Wickart et al. (2002).

31. Simulations with a CGE model for Switzerland show that in Switzerland AC pricing leads to a lower welfare loss in comparison with MSC pricing than in Belgium (Wickart et al., 2002). This can be explained by the lower congestion levels in Switzerland and by the fact that the ratio of transport revenue to financial costs is different in the two countries. In the reference equilibrium in Belgium revenue from the road transport modes is much higher than financial costs. In Switzerland revenue is approximately equal to financial costs. For public transport the rate of financial cost coverage is lower in Belgium.

REFERENCES

Calthrop, E. (2001). *On subsidising auto-commuting*. ETE Discussion Paper 2001–13 (www.econ.kuleuven.ac.be/ew/academic/energmil/publications.

De Borger, B. (2001). Discrete choice models and optimal two-part tariffs in the presence of externalities: Optimal taxation of cars. *Regional Science and Urban Economics*, *31*, 453–470.

Link, H., Stewart, H.L., Doll, C., Bickel, P., Schmid, S., Friedrich, R., Suter, S., Sommer, S., Marti, M., Maibach, M., Schreyer, C., & Peter, M. (2002). Pilot accounts: Results for Germany and Switzerland. *UNITE (UNIfication of accounts and marginal costs for Transport Efficiency)*. Working funded by 5th Framework RTD Programme, Deliverable 5, Summary Report. ITS, University of Leeds, Leeds.

Link, H., Stewart, L., Maibach, M., Sansom, T., & Nellthorp, J. (2000). The accounts approach. *UNITE Deliverable 2*. Funded by the 5th Framework RTD Programme. ITS, University of Leeds, Leeds.

Link, H., & Suter, S. (2001). *Summary of the German and Swiss pilot accounts – Background paper* – Draft version, UNITE Deliverable 5, Presented at Ecole Nationale des Ponts et Chaussées, Paris, 17–18 September 2001.

Mayeres, I. (1999). *The control of transport externalities: A general equilibrium analysis*. Ph.D. Dissertation, Faculty of Economics and Applied Economics, K.U.Leuven.

Mayeres, I. (2000). The efficiency effects of transport policies in the presence of externalities and distortionary taxes. *Journal of Transport Economics and Policy*, *34*, 233–260.

Mayeres, I., & Proost, S. (1997). Optimal tax and public investment rules for congestion type of externalities. *Scandinavian Journal of Economics*, *99*, 261–279.

Mayeres, I., & Proost, S. (2002). Testing alternative integration frameworks – Annex Report 1: The CGE model for Belgium. *UNITE (UNIfication of accounts and marginal costs for Transport Efficiency)*. Working Funded by 5th Framework RTD Programme, Annex Report 1 to Deliverable 13. ITS, University of Leeds, Leeds.

Mayeres, I., Proost, S., Quinet, E., Schwartz, D., & Sessa, C.C. (2001). Alternative frameworks for the integration of marginal costs and transport accounts. *UNITE (UNIfication of accounts and marginal costs for Transport Efficiency)*. Working Funded by 5th Framework RTD Programme, Deliverable 4. ITS, University of Leeds, Leeds.

Mayeres, I., & Van Dender, K. (2001). The external costs of transport. In: B. De Borger & S. Proost (Eds), *Reforming Transport Pricing in the European Union*. Cheltenham, UK and Northampton, MA, USA: Edward Elgar.

Mayeres, I. & Van Regemorter, D. (2003). *Modelling the health related benefits of environmental policies – A CGE analysis for the EU countries with GEM-E3*. ETE Discussion Paper 2003–10 (www.econ.kuleuven.ac.be/ew/academic/energmil/publications.

Parry, I. W. H., & Bento, A. (2001). Revenue recycling and the welfare effects of road pricing. *Scandinavian Journal of Economics*, *103*, 645–671.

Proost, S., & Van Dender, K. (2001a). The welfare impacts of alternative policies to address atmospheric pollution in urban road transport. *Regional Science and Urban Economics*, *31*, 383–412.

Proost, S., & Van Dender, K. (2001b). Methodology and structure of the urban model. In: B. De Borger & S. Proost (Eds), *Reforming transport pricing in the European union*. Cheltenham, UK and Northampton, MA, USA: Edward Elgar.

Proost, S., & Van Dender, K. (2004). Marginal social cost pricing for all transport modes and the effects of modal budget constraints. In: G. Santos (Ed.), *Road pricing: theory and practice*. Amsterdam: Elsevier Publishing.

Van Dender, K. (2003). Transport taxes with multiple trip purposes. *Scandinavian Journal of Economics, 105*, 295–310.

Wickart, M., Suter, S., & van Nieuwkoop, R. (2002). Testing alternative integration frameworks – Annex Report 2: Results from a CGE model application for Switzerland. *UNITE (UNIfication of accounts and marginal costs for Transport Efficiency)*. Working Funded by 5th Framework RTD Programme, Annex Report 2 to Deliverable 13, Version 0.1. ITS, University of Leeds, Leeds.

THE SOCIAL COSTS OF INTERMODAL FREIGHT TRANSPORT

Andrea Ricci and Ian Black

1. INTRODUCTION

1.1. The Policy Context

Freight intermodality is increasingly considered as a major potential contributor to solving the sustainability problems of the European transport sector. Many factors point at intermodal transport as a strategic option for Europe: road network limits (on many accounts), but also market globalisation, manufacturers seeking logistic rationalisation through the recourse to just-in-time and similar concepts, public authorities promoting a more efficient and sustainable use of land, growing generalised competition forcing the industry into the search for new profitability instruments, etc.

However, despite considerable investments in dedicated infrastructure, and the increasing awareness of the many benefits (economic, environmental) that a higher intermodal market share would generate, hardly 10% of the total volume of freight movements in Europe is carried out by intermodal options, while road transport still largely prevails.

Such a poor market performance points to an overall lack of competitiveness of intermodal transport services. To support current policy orientations, which consistently advocate a forceful promotion of intermodal transport, detailed assessments are therefore required.

Measuring the Marginal Social Cost of Transport
Research in Transportation Economics, Volume 14, 245–285
Copyright © 2005 by Elsevier Ltd.
All rights of reproduction in any form reserved
ISSN: 0739-8859/doi:10.1016/S0739-8859(05)14009-8

In fact, no single reason can fully explain the limited growth of intermodality so far, while many are commonly put forward, including:

- the inadequacy of the existing infrastructure, notably for what concerns the bottlenecks resulting from an insufficient integration in the planning of network links and nodes;
- the intrinsic complexity of the industry, which – as opposed to unimodal transport services – involves a wide variety of operators, calling for major organisational efforts in order to optimise the use of assets and resources; and
- the extensive role of the private sector and the lack of an appropriate integration platform, which, combined with the above-mentioned complexity, results in a highly unregulated, and often inefficient framework of operation.

While capacity expansion is likely to be crucial to the future growth of intermodal transport, it will by no means be sufficient to guarantee the desired modal shift.

Policies and actions must therefore be designed and implemented to:

- increase the productivity and efficiency of the intermodal sector (notably through technological and organisational enhancements); and
- reduce the imbalances currently observed between intermodal and road (notably through institutional, fiscal and pricing interventions).

The White Paper of the European Commission (EC) on the revision of the Common Transport Policy devotes special attention to intermodal freight transport services. In its section: "Linking up the modes of transport", it advocates a number of technical, economic and organisational innovations that directly aim at increasing the attractiveness of intermodal solutions. On the other hand, and no less importantly, many other measures and actions proposed by the White Paper, although they do not target intermodal freight transport as such, are immediately relevant to the general objective of promoting intermodality. Specifically one should mention: the revitalisation of European railways (through radical increases in efficiency and the eventual establishment of a dedicated freight network), the generalised improvement of the quality of transport services and, last but not least, the introduction of an adequate system of transport infrastructure charging.

1.2. Cost Information as a Basic Prerequisite

In such a context, policy formulation evidently requires an in-depth assessment of the current economic performance of intermodal freight transport, and more specifically a detailed knowledge of its costs and of the driving factors behind them. As shown in the diagram below, credible measurements of social costs (both internal and external) are in fact the common prerequisite for both public policy and private strategy formulation to enhance the competitiveness of intermodal freight services.

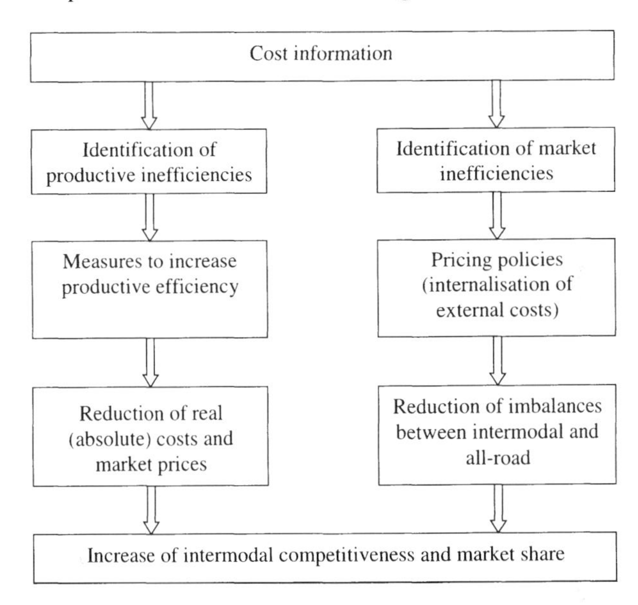

Ultimately, increasing the transparency of cost and price formation mechanisms stimulates fair competition, and, as a result, generates efficiency increases and the improvement in the quality of service. This in turn results in an increase of competitiveness for the European industry as a whole, and the strengthening of its position in the international playing field. Also, the diffusion of fair and efficient pricing practices is known to yield net welfare gains for the community at large.

In the above policy context, intermodal solutions should in fact be assessed as "just another transport mode", in order to: (i) investigate the credibility of the widely accepted view that intermodal transport is more efficient (in terms of total social costs) than its unimodal alternatives and (ii) provide the necessary input to price setting through reliable valuation analyses.

But intermodal transport is not a simple juxtaposition of modal activities, and the estimation of its costs requires additional original insights, e.g. into terminal and transhipment activities. At the outset, it is therefore necessary to identify the functional and technical specificity of intermodal chains, and the cost and price formation mechanisms attached thereto.

2. INTERMODAL FREIGHT TRANSPORT: ASSESSMENT OF CURRENT POSITION IN THE OVERALL EU MARKET

2.1. Definitions and Overall Market Position

Intermodal freight transport is defined as "the movement of goods in one loading unit (LU), which uses successively several modes of transport without handling of the goods themselves in transhipment between the modes".

This entails that:

- two or more different transport modes are deployed, and therefore at least one transhipment takes place; and
- the main haulage is not carried out by road, but by rail or water, while trucks/lorries are used for the initial and final legs of the goods movement (pre- and post-haulage).

The market share of intermodal transport in total European transport is limited: less than 10% of all intra EU15 transport (in tkm) takes place via intermodal transport. Nevertheless, all forms of European intermodal transport have shown a considerable growth over the last decade. Between 1990 and 1996, the average annual growth in tkm amounted to 9.3% for all forms of intermodal transport (European Commission, 2000). As for intermodal rail/road transport, however, volumes are reported to be declining since 1998, while short sea shipping and inland waterways intermodal transport are still growing fast.

2.2. Impacts of Institutional Factors on Supply (Costs) and Demand (Prices)

Both supply and demand of intermodal transport activities are directly influenced by the institutional setting in which the sector evolves. Legislation and agreements concerning, e.g. regulation and de-regulation, liberalisation and free market mechanisms are a significant part of this framework, which is in fact shaped at various institutional levels, namely European, national and subnational. While a detailed review of the relevant legislation does not find its place here, this section attempts to summarise some significant changes in the institutional setting that are bearing impacts on costs and prices of intermodal transport services.

Regulatory interventions affecting intermodal transport primarily focus on:

- the introduction of market forces, notably through the liberalisation of market access and the establishment of fair and efficient pricing systems within and between transport modes; and
- the establishment of a common legal framework, including environmental and social regulations, measures to promote combined transport and to regulate truck operations.

As for economic instruments, on the other hand, they mainly correspond to:

- funding, subsidies and incentives (e.g. through the European Development Fund (EDF), the Pilot Action for Combined Transport (PACT), the MARCO POLO programme, as well as transportation plans, policies and programs at the national and regional levels); and
- taxation and charges (e.g. vehicle taxation, fuel tax on diesel, the EUROVIGNETTE and its revision, motorway tolls).

In practice, Directives, Regulations and Agreements, providing for both normative and economic instruments, produce their effects on one or the other component of the intermodal chain (i.e. the various unimodal transport subsectors as well as terminal operations), which in turn translate into changes in the economic performances of the subsectors, e.g. through changes in the levels and structure of costs, and in market prices.

A direct, quantitative relationship between institutional interventions and economic performance is difficult to establish, owing to the great variety of concurrent measures undertaken at the various levels of government, and to the indirect nature of the causal link.

Nevertheless, several general conclusions can be reached:

- Liberalisation has certainly contributed to a decrease in market prices on most accounts. Although a thorough comparison based on actual price levels is difficult to carry out, a rough estimate indicates that prices have dropped approximately 30/40% as a consequence of the various liberalisation interventions.
- On the other hand, the regulatory regime has encouraged haulage firms to compete by providing supplementary services, which has resulted in their

carrying more goods and providing more services than have been priced and reported to the regulation authority. Balancing the price and the service effects, an estimate of a net price reduction of between 15% and 25% appears reasonable.

- It is however in the area of social regulations that the main influence on performance should be sought: limits imposed on allowed driving time, for instance, have a direct and substantial impact on costs and prices, considering that driver costs account for some 50% of all road operation costs.
- With reference to inland waterway transport, the harmonisation of transport subsidisation policy, which began with Directive 70/1107, has helped slow down the use of subsidies in the EU member countries and stabilise the market for inland waterway shipping.
- A main market distortion – often evoked by competing sectors – is to be found in the Short Sea Shipping sector, where navigation industry finances only a minor share of infrastructure costs which, depending on cost accounting methods, is estimated between 10% and 15%.

3. COST AND PRICE FORMATION ALONG THE INTERMODAL CHAIN

The general layout of an intermodal chain can be described as a sequence of 11 main activities, each generating internal and external costs: the activities are those by the shipper in loading a consignment, pre-haulage to a terminal (in which a transhipment point in the form of storage can be inserted), the terminal handling near to the shipper, the main haulage (with train, truck, ship or inland waterway), the terminal handling near to the consignee, the post-haulage and finally the consignee receiving the consignment. Within these 11 activities, there is no difference between the cost components of the shipper and a consignee and no difference between those of pre- and post-haulage. This means that the cost classification can refer to nine activities each with its own set of internal and external cost components.

3.1. Loading/Unloading – Shipper/Consignee

The first and the last step of a transport process involves the companies which dispatch and receive the consignment. Their costs include those incurred in loading (unloading) and storing the units used for transport (containers, trailers). In addition, a company may incur costs in leasing or owning units used for transporting the goods. External costs are geared to the use of machinery and equipment for the above operations.

3.2. Pre-Haulage/Post-Haulage

Pre- and post-haulage to and from terminals (e.g. rail) is typically provided by road transport companies. Road haulage companies incur costs involved in the ownership and operation of vehicles, which in most cases involve the payment of taxes. The total costs include the time spent loading and unloading as well as movement. Costs may also be incurred for the payment of infrastructure in the form of tolls. External costs are generated by the truck movement.

3.3. Transhipment

This can be defined as the location in which LUs are physically transhipped from one vehicle to another similar vehicle of the same mode (e.g. truck to truck). Internal costs involved are the capital cost of the equipment necessary for transhipment, its operation and the storage area required. External costs are geared to the use of machinery and equipment for the above operations.

3.4. Terminal Transfer

A terminal is defined as a place containing the functions and technical assets whereby an LU may be transhipped between two *different* kinds of carrying units. Transfer may be between the various modes – rail, road, sea and inland waterway. External costs are geared to the use of machinery and equipment for the above operations.

3.5. Marshalling Yard Transfer

The function of a marshalling yard is the transhipment of LUs from one *train* to *another*, or, more commonly, the rearrangement of wagons into a single train. External costs are generated by the movement of vehicles (locomotives, wagons) and by the use of other equipment and machinery for the transfer operations.

3.6. Main Haulage: Road

Road haulage companies incur costs involved in the ownership and operation of vehicles, which in most cases involves the payment of taxes. Costs may also be incurred for the payment of infrastructure in the form of tolls and road pricing. This includes all national road tax stickers and motorway vignettes operative in Switzerland and Austria as well as tolls for Alpine and Channel crossing. External costs are generated by the movement of trucks.

3.7. Main Haulage: Rail/Train

The costs refer to a *terminal*-to-*terminal* journey performed by rail transport. They include any charges for the use of the rail infrastructure. These charges may or may not cover the costs of the infrastructure. External costs are generated by the movement of trains.

3.8. Main Haulage: Inland Waterway

The cost structure of this transport block is similar to that experienced for main haulage by train. Charges may be incurred for the use of infrastructure. External costs are generated by the movement of barges.

3.9. Main Haulage: Maritime

The cost structure of this transport block is similar to that experienced for main haulage by train. No charge is paid for infrastructure (the sea). Costs for piers and berths maintenance and repair are normally allocated to terminals. External costs are generated by the movement of ships.

Deriving *unit costs and prices* that are relevant to both policy makers and market players is a complex affair:

- The market for intermodal transport is organised with reference to the movement of LUs, rather than vehicles. In fact, the LU can be considered as the equivalent of the vehicle in other forms of transport, whereby it is the LU – and not the various vehicles on which it is successively loaded – that physically moves all the way from origin to destination. But LUs can be of different types, with three main options: containers, swap bodies and trailers, each with its characteristics that relate to different organisational, technical and market needs. Despite the current trend towards standardisation, several solutions remain therefore available, even within each LU category, e.g. containers that can be 20- or 40-ft long. For pricing purposes, analysts and operators alike therefore tend to express values in Euro/LU km rather than in Euro/veh km, as the former are more market-relevant and of direct interest to end-users (shippers and consignees).

- On the other hand, transport charging and taxation policies commonly refer to values in Euro/veh km, therefore prompting the need for a "translation" of values expressed in LUs to values expressed in vehicle terms. Such conversion is far from straightforward, if one considers that a given vehicle (e.g. a 40 tonnes articulated truck) can accommodate various combinations of different LUs, with varying load factors.

4. MEASURING THE MARGINAL SOCIAL COSTS OF INTERMODAL TRANSPORT

4.1. An Integrated Accounting Framework for Intermodal Freight Transport

A detailed accounting framework for intermodal freight transport has been devised within the RECORDIT project (RECORDIT), and is summarised in Table 1 below. It covers the full range of intermodal activities and allows the representation of both internal and external costs, using the following definitions.

Internal Costs are those incurred by an operator responsible for one of the eleven activities. They are broken down into seven major components (each with a number of sub-components):

1. Personnel
2. Fixed assets/maintenance of assets
3. Energy, other consumption materials/telephone, telecommunication and radio
4. Stock turn
5. Time

Table 1. Cost Components and Activities.

	Shipper/ Consignee	Pre- and post-haulage	Transship Point	Terminal	Marshalling	Haulage Road	Haulage Rail	Haulage Waterway	Haulage Maritime
Internal Costs									
Personnel									
Gross wage/salary									
- Of driver		X			X	X	X	X	X
– Of worker	X		X	X	X				
Expenses incurred by the driver		X				X	X	X	X
Social security	X	X	X	X	X	X	X	X	X
Overhead	X	X	X	X	X	X	X	X	X
Administration	X	X	X	X	X	X	X	X	X
Profit/opportunity	X	X	X	X	X	X	X	X	X
Advertising, PR	X	X	X	X	X	X	X	X	X
Advocating/Consulting	X	X	X	X	X	X	X	X	X
Fixed assets/maintenance of assets									
Container/swap body/trailer									
Investment: depreciation and interest	X	X	X	X		X	X	X	X
Maintenance	X	X	X	X		X	X	X	X
Repair	X	X	X	X		X	X	X	X
Means of transport									
Inv.: dep. and interest		X				X	X	X	X
Maintenance		X				X	X	X	X
Repair		X				X	X	X	X
Technical asset									
-Inv.: depreciation and interest			X	X	X				
Maintenance and repair			X	X	X				
Building									
– Inv.: depreciation and interest	X		X	X	X				
– Maintenance and repair	X		X	X	X				
Property/site/development									
– Inv.	X		X	X	X				

Infrastructure									
– Inv.: depreciation and interest		X	X	X	X	X	X		
– Maintenance and repair		X	X	X	X	X	X	X	X
Energy, other consumption materials/ telephone, telecommunication and radio									
Fuel, diesel		X			X	X	X	X	X
Electricity	X		X	X	X		X		
Oil, fat, additional variable cost		X	X	X	X	X	X	X	X
Tyres		X				X			
Telephone, telecommunication, radio	X	X	X	X	X	X	X	X	X
Stock turn									
Loading/Unloading	X								
Transhipment			X	X					
Shunting, marshalling, rearrangement			X		X				
Storage of goods	X		X	X					
Time									
Waiting time		X	X	X	X				
Rest time for driver						X			
Parking, port liner terms charge			X			X		X	X
Organisation costs									
Monitoring	X	X	X	X	X	X	X	X	X
Safety test	X	X	X	X	X	X	X	X	X
Disposition of wagon/vehicle fleet									
Additional keeping ready of wagons and means of transport	X	X	X	X	X	X	X	X	X
Disposition of cargo/good									
– Dispatching, conducting, co-ordination	X								
Operational cost for the network (rail/ waterway-signalling, station and network management)			X		X		X	X	X
Management/Transaction	X	X	X	X	X	X	X	X	X

Table 1. *(Continued)*

	Shipper/ Consignee	Pre- and post- haulage	Transship Point	Terminal	Marshalling	Haulage Road	Haulage Rail	Haulage Waterway	Haulage Maritime
Insurance/taxes/charges									
Insurance									
– Of cargo/good	X								
– For the risk of the enterprise	X								
– For vehicle and LUs		X	X	X	X	X	X	X	X
Third party motor vehicle insurance		X				X		X	X
Tax, sales tax	X								
Vehicle tax		X				X			
Duty	X								
Tolls, road-pricing		X				X			
Fixed road charges, truck vignette		X				X			
Rail track user charges					X		X		
Lock charge								X	
Costs with internal and external parts									
Congestion		X				X		X	
Scarcity, slot allocation			X	X	X		X	X	X
Specific road bottleneck, go round						X			
External costs									
Accident	X	X	X	X	X	X	X	X	X
Air pollution	X	X	X	X	X	X	X	X	X
Climate change	X	X	X	X	X	X	X	X	X
Noise nuisance	X	X	X	X	X	X	X	X	X

6. Organisation costs
7. Insurance /taxes and charges

External costs are those incurred by other parties as a result of an operators transport or terminal activities. These comprise costs stemming from accidents, air pollution, climate change and noise nuisance.

An intermediate category, referred to as ***costs with internal and external parts***, identifies those costs where an operator incurs internal costs as a result of an activity and also imposes costs on other parties. The obvious example here is congestion.

The price formation mechanism is in turn illustrated in Fig. 1 below, along the following definitions:

A *charge* is defined as the price that is paid to infrastructure operators i.e. the road, rail and inland waterway operator, respectively. These are paid by road haulage, marshalling yards, train operators and inland waterway operators who incur them as costs.

Price refers to the income received by an operator for a transport or terminal movement. The difference between total revenue and total cost is

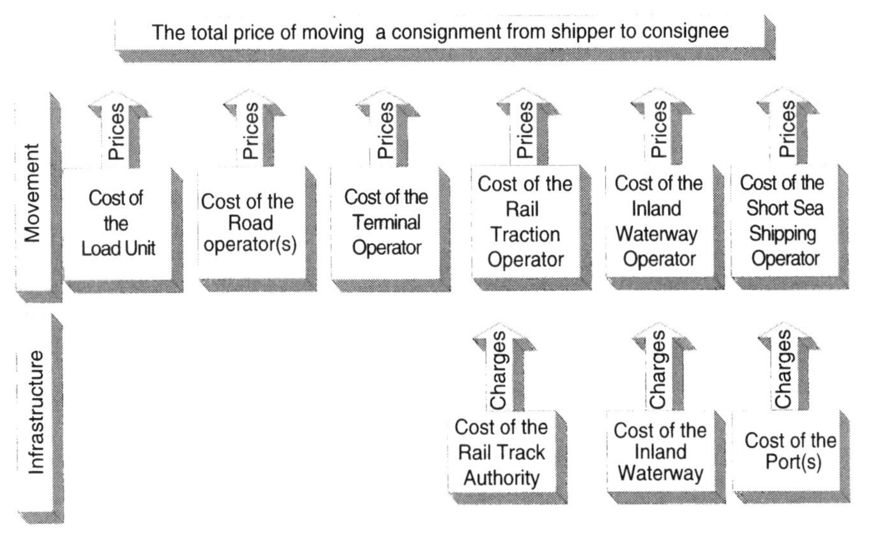

Fig. 1. Charges, Costs and Prices in Intermodal Transport.

profit or loss; and between price and unit cost it is profit per unit (load unit moved). From the perspective of the purchaser of transport services, prices in the market become his costs.

4.2. Cost Drivers and Cost Functions

Ideally, the calculation of marginal social costs requires the identification of cost functions integrating all factors that influence each of the cost items described through the corresponding functional relationships. In practice, such cost functions are hardly identifiable in the general case, and cost measurement mostly relies on empirical observations. There are however a number of factors that influence unit costs and which may vary for a particular activity:

1. the unit costs of resources
2. the efficiency of operations
3. the load factors
4. the length of a journey leg
5. the characteristics of the journey leg
6. the scale of movement

Remarkably, all such cost drivers influence both internal and external costs.

4.2.1. Unit Costs of Resources Vary throughout Europe

The unit cost of personnel (drivers, maintenance, staff) varies between different countries. The cost of materials may also vary. Finally, total internal costs differ because taxes, charges and subsidies vary throughout the Union.

4.2.2. Efficiency of Operations

The provision of a transport or terminal movement requires factors of production or resources. The combination of resources used to provide a given service (the production function) may vary in different locations for a variety of reasons. Some operations may include more personnel on the vehicles, maintenance may be carried out more frequently and fuel consumption may vary between different vehicles. All these factors lead to differences in the cost per unit capacity provided. Another element of efficiency is the intensity of use of vehicles (in all modes). To some extent capital costs can be considered fixed in a period of time (a year); hence the greater the use of that vehicle, the lower the unit cost.

4.2.3. Load Factors

All transport modes are characterised by load factors on vehicles significantly below 100%. (Road freight vehicles typically travel empty for 30% of the time and when loaded have a load factor of 70% (European Commission, 1999, REDEFINE). The level of utilisation achieved in operations therefore has a large impact on the unit costs (per load unit carried kilometre). In turn, the load factor varies due to a number of reasons:

The imbalance of two way flows – where this is high, the load factor in one direction will certainly be low and thereby depress the overall average.

Length of journey – load factors tend to be high on long journeys by road, where operators have a strong incentive to find return loads and consolidate loads. This feature leads to distinct differences in pre-haulage and trunk haulage load factors and hence unit costs.

4.2.4. The Length of a Journey Leg

It is well known (European Commission, 2000, SOFTICE) that unit prices for freight transport tend to be higher for short journeys. This phenomenon is based largely on the underlying cost function (as well as the issue of changing load factors mentioned above). The reason for this stems from the costs at the beginning and end of a journey (concerned with aspects of loading/unloading), which are incurred regardless of length (and thus comprise a semi-fixed cost).

4.2.5. The Characteristics of the Journey Leg

The most important component is the speed of the journey. Some cost components are related to distance (fuel consumption, maintenance); others are related to time (drivers wages, vehicle depreciation). Fuel consumption in particular is also directly influenced by speed. Each cost component is therefore related to either distance or time (speed). No general agreement exists on how to describe these dependencies: capital cost of vehicles is sometimes assumed to be time based (constant depreciation per year) and sometimes distance based (per kilometre). The realistic answer is probably a combination. Average speed will therefore be a strong determinant of unit costs for road, rail, maritime and inland water movements.

4.2.6. The Scale of Movement

As the flow of load units increases along a journey leg or through a terminal unit, costs may decrease or increase. In the former case we speak of economies of scale and in the latter diseconomies. In the former case, the marginal cost of a load unit is below average cost (and vice versa for diseconomies of scale).

There is some flexibility about defining marginal costs concerned with temporalities and integer effects. The critical question is how the use of factors of production is assumed to respond to a change in production. An extra load unit carried between A and B might only require extra fuel (assuming the driver and vehicle are already available). Or, even if a new driver and vehicle are required, there may be no associated increase in the overheads (management) of the business. The response may be different over time – some factors may only respond after a number of months. Transport vehicles often carry a large number of load units, and it can be argued that an extra load unit can be carried for a marginal cost near to zero (until of course capacity is reached). This problem (due to capacity being provided in integers) is even more acute where there are items of capital equipment, which provide large elements of capacity. Infrastructure in the form of track and some terminal equipment provides the obvious example. All of these considerations are related to the possible existence of spare capacity and the possibility that some costs can be regarded as fixed for some range of output.

In the case of haulage, another element associated with design may affect costs. Where demand is low, an operator may use small vehicles steadily increasing the size as demand increases. Large vehicles carry loads more cheaply (either on a per load unit or per tonne basis) than smaller vehicles (in whatever mode). This phenomenon (essentially economies of scale in vehicle size) therefore can lead to unit costs falling, as demand increases.

The relationship between these different elements can be seen in the following diagram (Fig. 2).

Scale of operations has different impacts along the intermodal chain:

- *Shipper/consignee*. The activities associated with handling inbound and outbound load units are likely to be strongly related to the number of units. There are no factors associated with fixed capital equipment that should lead to economies of scale. Marginal costs should therefore be the same as average costs.
- *Pre- and post-haulage. Main haulage road*. A sustained increase in demand in the absence of spare capacity requires a *pro rata* increase in all inputs (vehicles, drivers, management). If there is no increase in the load factor (and there is no evidence that it does change with scale), returns to scale are constant.
- *Main haulage train*. As for operational costs associated train haulage, it seems reasonable to ignore the integer effect of spare capacity on

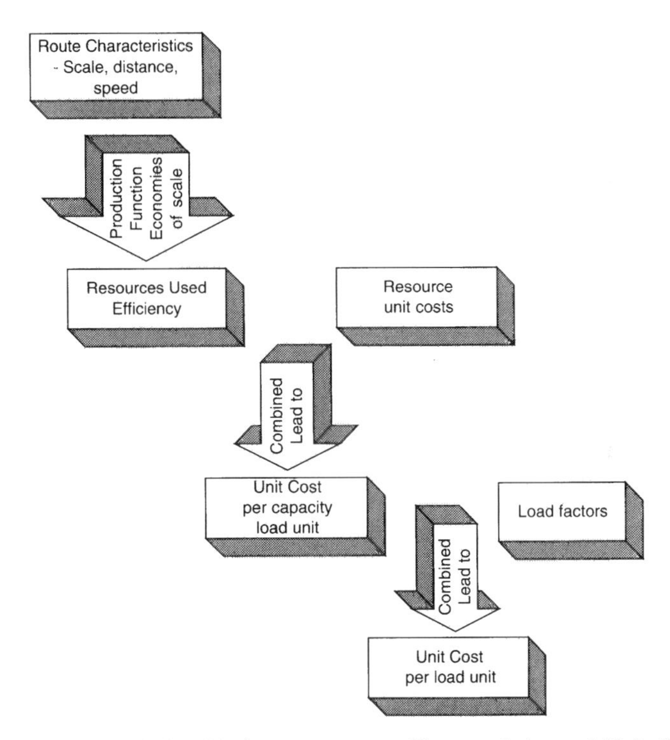

Fig. 2. The Relationship between Route Characteristics and Unit Cost.

individual trains and assume an increase in demand will bring forward a *pro rata* increase in the number of trains required. This means the unit costs associated with haulage remain constant. On some routes with low demand, where an increase in demand could justify the use of larger vehicles, this would lead to lower unit costs. As far as management costs are concerned, it is probably legitimate to assume that an increase in demand will require a comparable increase in management effort. From an operators perspective, the cost components associated with track charges might be fixed (annual payment) or, more commonly, related to train kilometre. The latter case implies constant unit costs. Overall therefore, assuming that scale does not affect the load factor, there are likely to be constant returns to scale in train haulage.

- *Main haulage inland waterway*. Similar arguments to those presented for rail also apply in this activity; and therefore in terms of operations constant returns to scale will apply, unless factors of production are identified that do not change *pro rata* with scale.

- *Main haulage maritime.* Maritime differs from rail and inland waterways in that a wide variety of vessel sizes are in use, implying that demand increases may allow the use of larger vessels and hence the enjoyment of some economies of scale. Vessel handling costs also exhibit some element of scale economies.
- *Transhipment point, terminal, marshalling yard.* These activities are deemed to experience economies of scale over some ranges of throughput (and hence a range where marginal costs are lower than average costs), owing to the heavy capital investment in yard facilities concerned with railway line provision, areas for loading and large capacity cranes. These components tend to comprise a fixed cost over a wide range of throughput. Scale economies are particularly noticeable where throughput is low. At higher levels, costs are likely to contain strong discontinuities when a new piece of capital equipment is required.

Two additional issues should also be mentioned with respect to economies of scale:

- *Track costs.* In the case of both rail and inland waterways (using canals), track costs exhibit strong economies of scale. Variable costs "... *are likely to account for a small proportion (typically between 10% and 20%) of total rail infrastructure costs. Other costs are unlikely to be affected by changes in the volume of traffic, for example some maintenance and renewals costs will be driven by time, and also by the mix of traffic using a route, rather than the number of trains using the network*". This report also brings attention to the other elements of cost that should be included when dealing with infrastructure. "*When a network becomes congested, then further potentially very significant costs are included... Initially, these cover the costs of an increased risk of disruption, since a delay is far more likely to cause disruption to other services on a congested route... At even higher levels of capacity utilisation, infrastructure managers will find that they cannot satisfy train operators demands for access. ... In each case, the opportunity cost (i.e. the net revenue foregone by the infrastructure manager) should be included in the ... (marginal cost)*".
- *Mohring effect.* It has been observed (usually with reference to passenger transport) that increased demand leads to a greater frequency of service. This phenomenon (usually referred to as the Mohring effect) is applicable to intermodal transport. As demand increases following for example a relative improvement in intermodal costs vis-à-vis road, the improved frequency may generate a further increase in demand.

4.3. The RECORDIT Corridors

This section presents selected results of marginal social cost measurements carried out within RECORDIT. It covers both internal and external costs associated to the movement of LUs along three door-to-door pan-European corridors:

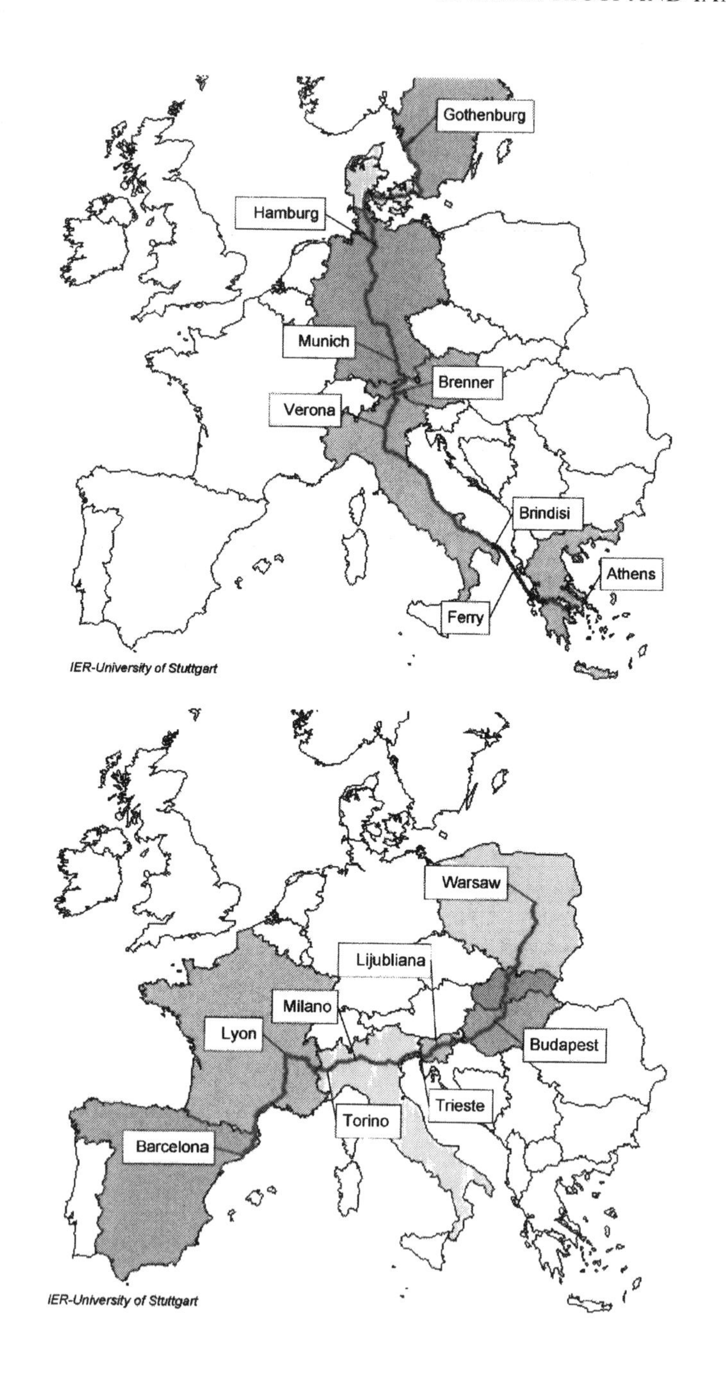
Gothenburg
Hamburg
Munich
Brenner
Verona
Brindisi
Athens
Ferry
IER-University of Stuttgart
Warsaw
Lijubliana
Milano
Lyon
Budapest
Trieste
Torino
Barcelona
IER-University of Stuttgart

Consistently with the main objective driving the promotion of intermodal transport services, that is to achieve significant modal shifts from all-road to intermodal solutions, the cost calculations have systematically been carried out in parallel for both options, and are presented so as to facilitate the comparison of economic performance between the two modes.

In choosing the RECORDIT corridors, great care was taken to identify routes that would prove representative of the overall European market of intermodal transport services, thereby adequately lending themselves to a possible generalisation process.

Clearly, both the sheer length and the geographical and modal variety of the selected corridors reflect this objective, with corridors that cumulatively extend over 16 countries, more than 9,000 km of routes, and all transport modes (with the exception of air).

One should however be cautious in attempting direct extrapolations from the corridor results, for a variety of reasons:

- RECORDIT deals with door-to-door transport solutions. The market situation is however such that, on any of the RECORDIT corridors, the current door-to-door transport demand (i.e. LUs being moved all the way from Patras to Gothenburg, etc.) is in fact very limited. While the selected routes are theoretically sound – as they allow for a wide coverage of modes and contexts – the results (in terms, for example, of unit door-to-door costs) are not immediately market-relevant.
- Also, estimated costs (both internal and external) vary considerably from one corridor to another, and approaches based on the adoption of average values raise important issues in a generalisation perspective.
- Moreover, uncertainties affecting individual cost values can be very high, owing in particular to: (i) the difficulty in obtaining high quality bottom-up data, (ii) the difficulty in establishing common sets of assumptions across countries and corridors and (iii) the intrinsic uncertainty associated with valuation methodologies.

On the other hand, several steps have been taken to – at least partially – offset these limitations:

- While door-to-door traffic is indeed scarce on the full length of the RECORDIT corridors, intermodal services are offered and active on most sub-segments therein. This allows one to assess shorter, market-relevant corridors as an immediate by-product of the full corridor analysis.
- RECORDIT has established a comprehensive database with all the values of the individual cost items assessed along the corridors and sub-corridors.

Ultimately, this wealth of highly disaggregated data (several hundreds of individual cost items) provides an extensive and (to a large degree) statistically significant sample of elementary observations, which lends an acceptable level of credibility to further quantitative analyses carried out in a generalisation perspective.
- Finally and most importantly, RECORDIT has developed software (the RECORDIT DSS) to allow for a flexible and multi-purpose use of the cost data assembled through corridor case studies. Sensitivity analyses can be easily carried out to partially compensate for uncertainties; new corridors can be defined and assessed by re-assembling building blocks from existing routes and the cost values thereof; policy-induced changes in the value of individual cost items (e.g. the elementary cost of a crane movement in a terminal, the fuel consumption of a truck, etc.), or on cost categories (e.g. cost of labour, fuel taxes, etc.) can be easily simulated to produce new estimates of corridor costs, at all possible disaggregation levels.

Each of the RECORDIT corridors is defined in terms of the beginning and end of the intermodal points on a route. The cost analysis is carried out for a door-to-door movement between an origin and a destination in the vicinity of these locations. The three intermodal routes in the corridors therefore require pre-haul from the origin and post-haul to a destination. In the first corridor, for instance, the post-haul is from Manchester to Preston (a distance of 50 km) and in the second there is a pre-haul from Athens (a distance of 210 km). Each corridor is served by an intermodal route and an all-road route.

4.4. Internal Costs

4.4.1. Methodology

The calculation of internal costs along the corridors follows the cost classification specified in the reference accounting framework, and therefore entails a very disaggregated breakdown, with some 800 cost items documented. Broadly speaking, internal costs belong to six main cost categories, namely (i) staff costs, (ii) fixed assets and maintenance (excluding transport infrastructure), (iii) energy and other consumptions, (iv) stock turn, (v) organisation and management costs and (vi) insurance, taxes and charges. Transport infrastructure costs (maintenance, wear and tear) are not explicitly included inasmuch as they are not directly paid the infrastructure users, although they might be totally or partially reflected in taxes and charges.

The primary data source for internal costs is direct observation, through surveys carried out with the involved players (unimodal operators and infrastructure managers, multimodal operators and shippers).

Supplementary information elicited from available literature and company accounts allows one to fill gaps and ensure overall data consistency.

4.4.2. Cost Values

Table 2 summarises costs per LU and costs per kilometre (including taxes in both cases) for the movement of a 40-ft container along the three corridors.

The total internal cost of intermodal can be further broken down into its major components of which the most important are pre-/post-haulage, the haul leg (by different modes) and transhipment costs. Fig. 3 shows this breakdown for the three corridors.

Table 2. Comparison of Internal Costs between Intermodal and All-road (1998).

Corridor	Intermodal			All-Road		
	€/LU	Length (km)	€/km	€/LU	Length (km)	€/km
Genova–Preston	2315	2134	1.08	2836	1912	1.48
Athens–Gothenburg	3970	4128	0.96	4894	3599	1.36
Barcelona–Warsaw	3350	3270	1.02	3448	2735	1.26

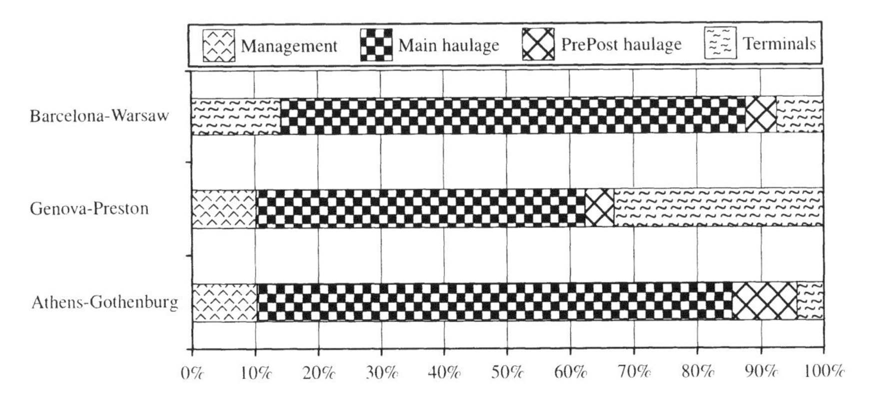

Fig. 3. Breakdown of Intermodal Costs.

While main haulage is the most important cost in all cases, the share of movement and transhipment at terminals can reach and in some cases exceed 20%. The shares vary depending critically on the number of transhipments necessary along the intermodal route, the length of the pre- and post-haul legs and the route distance. In routes less than 1,000 km, for instance, the share of main haulage is likely to be less than 85% and may fall as low as 50% for the shortest routes.

4.5. External Costs

4.5.1. Methodology

For environmental externalities, RECORDIT has adopted the damage cost approach, based on the Impact Pathway methodology, which starts from the technical characteristics of the activity (technology and type of vehicle, load factor, corridor length), then calculates the so-called "burdens" associated with the activity (i.e. emissions of pollutants, emissions of noise, frequency of accidents), then models the physical impact of these burdens on human health, crops, materials, etc., and finally estimates the monetary value of these damages (through market values when available, as with crops and materials, or through Willingness-to-Pay values otherwise). This process – which, as described, is strictly bottom-up – was originally devised for and used in the ExternE project, and has since been used and validated in several other projects and case studies (e.g. QUITS, UNITE); although uncertainties still remain, it is generally considered to be accurate (see Chapter 7 for a detailed discussion of this methodology).

4.5.2. Cost Values

This section presents a summary of the results of the corridor analysis of external costs in intermodal transport. Table 3 illustrates the comparison between intermodal transport and the all-road option per LU, i.e. for the movement of a 40-ft container along the three corridors.

Unit external costs for rail show minimal variations across segments and corridors. As for road pre- and post-haulage, the high cost shown for Barcelona is mainly due to the high accident rate in Spanish urban agglomerations. Altogether, the external costs of intermodal transport are 50–70% lower than for all-road.

Figs. 4–6 show the disaggregation by external cost category.

Not surprisingly, accident costs account for a major share of the difference between the two options, with the partial exception of the

Table 3. Marginal External Costs: Comparison between Intermodal and All-road.

Segment	Mode	€	€/km
Genova to Preston			
Genova–Basel	Rail	30	0.06
Basel–Rotterdam	IWW	73	0.04
Rotterdam–Felixstowe	SSS	29	0.14
Felixstowe–Manchester	Rail	18	0.04
Manchester–Preston	Post-haulage (road)	16	0.34
Transhipments		1	
Genova–Preston	Intermodal	167	0.08
Genova–Preston	All-road	448	0.24
Athens to Gothenburg			
Athens–Patras	Pre-haulage (road)	80	0.38
Patras–Brindisi	SSS	263	0.59
Brindisi–Gothenburg	Rail	160	0.05
Gothenburg	Post-haulage (road)	39	0.78
Transhipments		13	
Athens–Gothenburg	Intermodal	586	0.15
Athens–Gothenburg	All-road	1122	0.31
Barcelona to Warsaw			
Barcelona	Pre-haulage (road)	135	2.70
Barcelona–Warsaw	Rail	205	0.06
Warsaw	Post-haulage (road)	43	0.86
Transhipments		1	
Barcelona–Warsaw	Intermodal	384	0.11
Barcelona–Warsaw	All-road	917	0.34

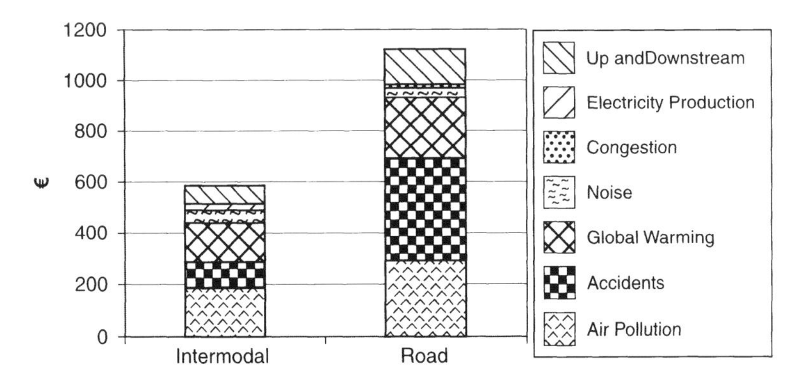

Fig. 4. External Costs – Athens to Gothenburg.

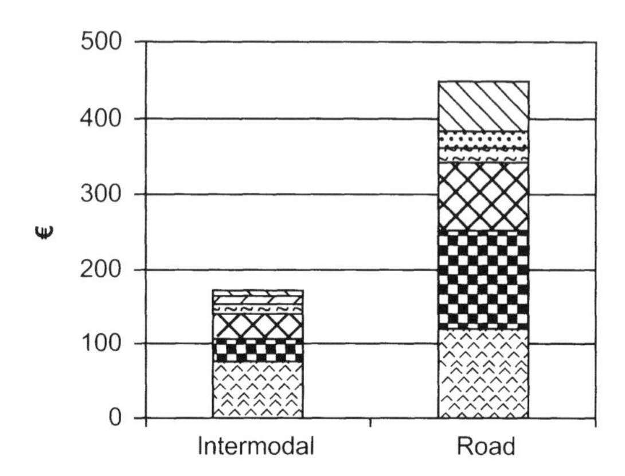

Fig. 5. External Costs – Barcelona–Warsaw.

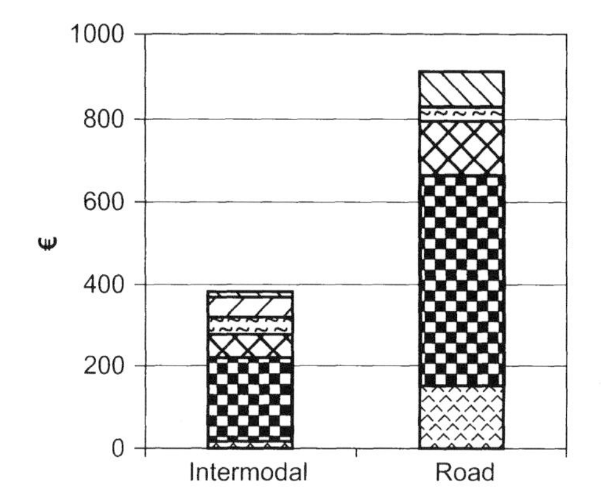

Fig. 6. External Costs – Genova–Preston.

Barcelona–Warsaw intermodal corridor, which suffers from the high incidence of urban accidents on the pre-haulage leg. Electricity production costs (rail) only partially compensate for the high air pollution costs of road transport, while global warming is consistently lower for intermodal transport in all corridors. Congestion costs are only significant for the Genova–Preston corridor, mainly owing to the dense agglomeration area crossed by this corridor between Milano and the Swiss border at Chiasso.

5. MARGINAL SOCIAL COSTS AND POLICY FORMULATION IN THE INTERMODAL TRANSPORT SECTOR

5.1. Assessing Market Imbalances: Current Cost Coverage of Competing Options

It is commonly thought that road transport enjoys a competitive advantage with respect to intermodal competition owing to the limited degree of internalisation of its external costs, which are in general higher than for intermodal solutions. While intuitively reasonable, this belief should be validated through scientific measurements, if only to assess the extent of the current imbalance, and to identify the specific areas where it occurs: contrary to unimodal transport, intermodal services entail a complex sequence of production steps, including invariably recourse to road transport for the pre- and post-haulage legs. Moreover, it is common knowledge that the trucks used for pre- and post-haulages are often of a poor technological standard, leading to low environmental performance. Also, intermodal chains include a variety of terminal operations that do not appear in all-road transport (transhipment at terminals, marshalling yards), whose external costs must be added to those of the haulage proper. Finally, trucks are subject to charges and tolls: irrespective of the actual use that is made of the corresponding revenues, and what matters when assessing the level of competitiveness is the impact that these disbursements have on the modal choice of users.

To assess the existence and the extent of these modal imbalances, a systematic comparison must be carried out between the full costs of infrastructure use and those currently paid by the user.

Whether or not transport users pay for the marginal social costs they generate can be assessed through the following simple calculation:

$$\text{Social costs balance (SCB)} = \text{External costs} - \text{Taxes} - \text{Net infrastructure payment}$$

where

$$\text{Net infrastructure payment} = \text{Infrastructure charges and tolls} - \text{Marginal costs of infrastructure use}$$

A major prerequisite is then the measurement of the actual infrastructure costs imposed during a journey. In the perspective of marginal social cost accounting, infrastructure costs only refer to the cost of maintenance and operations imposed by a vehicle movement. The capital cost of the infrastructure is regarded as a fixed cost and is not included. Marginal costs of road and rail infrastructure are hardly measurable at the level of specific routes, and average values are then used to reflect marginal costs at network level. There still remains considerable disagreement over the method that should be used to calculate marginal infrastructure costs both in road and rail, and therefore substantial uncertainty is attached to these reference figures (Link, Dodgson, Maibach, & Herry, 1999).

Charges and tolls paid for the use of road and rail infrastructure are generally known at route level. Table 4 shows the results of SCB calculation on the three RECORDIT corridors, for the movement of a 40-ft container. Positive values of SCB indicate a current underpayment of marginal social costs.

Based on the above calculations, it appears that all-road transport indeed enjoys a competitive advantage on two of the three routes, i.e. Genova–Preston and Barcelona–Warsaw, although on the latter it fails to cover its own social costs.

In the case of Athens–Gothenburg, on the other hand, the situation appears reversed, whereby the intermodal solution fails to cover its external costs (the major part of which is caused by the pre- and post-haul legs by road) while the road solution shows an approximate balance.

On the Barcelona–Warsaw route, the charge that would be required to cover marginal social costs on the road system is €0.16/40′ container. The difference from the other corridors is that external costs are higher (again a large part of the explanation is different accident costs) and most critically that taxes paid are very low. The use of trucks from Croatia to carry the goods for a large part of the journey means that fuel and circulation taxes are very low for these segments. At the same time, however, the charge required for intermodal to cover its external costs is also substantial (at €0.11/40′ container km).

5.2. Increasing the Competitiveness of Intermodal Transport through the Reduction of Costs

Cost reductions usually target improvements in productive efficiency, and thus increases in profitability. As previously highlighted, however, a decrease in the costs of transport service production may as well, depending on the type of

Table 4. Results of the SCB Calculation on the Three RECORDIT Corridors.

		€	€/km
Genova to Preston			
Intermodal	External costs	167	0.08
	Taxes paid	284	0.13
	Net infrastructure payment	−26	−0.01
	SCB	−91	−0.04
Road	External costs	448	0.24
	Taxes paid	404	0.21
	Net infrastructure payment	14	0.01
	SCB	30	0.02
Athens to Gothenburg			
Intermodal	External costs	586	0.14
	Taxes paid	140	0.03
	Net infrastructure payment	183	0.04
	SCB	263	0.06
Road	External costs	1122	0.31
	Taxes paid	729	0.20
	Net infrastructure payment	460	0.13
	SCB	−67	−0.02
Barcelona to Warsaw			
Intermodal	External costs	384	0.11
	Taxes paid	66	0.02
	Net infrastructure payment	−48	−0.01
	SCB	366	0.11
Road	External costs	917	0.34
	Taxes paid	226	0.08
	Net infrastructure payment	250	0.09
	SCB	441	0.16

intervention, bring along a reduction of external costs, thereby also enhancing the social attractiveness of the service. This section presents three exemplary scenarios of intermodal transport cost reduction. All three are directly geared to explicit policy priorities identified by the EU (e.g. in the 2001 White Paper).

5.2.1. Rail Interoperability and Efficiency Improvements in Terminals/ Transhipment Operations

The aim here is to explore the overall cost reduction when three different initiatives are applied to terminals and the transhipment process. These three initiatives apply to operations, manpower management and infrastructure management.

(1) *Eliminate rail–rail transhipments.* It is clear, particularly from the example of the Barcelona–Warsaw route (6 vertical transhipments) that a significant cost derives from an excessive number of transhipments between rail and rail (the interoperability issue). Some of these transfers involve large marshalling yards and transfer from one train to another. The cost reduction scenario assumes that all of this type of transfer are eliminated. Clearly, this represents a change only achievable in the medium to long term.
(2) *Improve manpower productivity.* Empirical evidence is available (e.g. from RECORDIT) that management of the labour force at some terminals leads to lower productivity than desirable. Accordingly, productivity increases at terminals is explicitly targeted by the 2001 White Paper. It is here assumed that increased manpower efficiency could, in the short/medium term, lead to a reduction of 15% in labour costs per LU through all terminals (ports, road/rail, road/barge etc.).
(3) *Reduced capital costs.* There is some potential in the medium term for a reduction in the fixed costs of terminals through the use of more efficient layout, more appropriate equipment and more rational use of capacity. An upper value of 33% fixed cost reduction per LU has been used here for a change that could occur in the medium to long term.

The effect of this scenario in the three RECORDIT corridors is an average reduction in resource costs of around 6%. Simulation suggests that on shorter routes such cost reduction decreases, and may ultimately not exceed 1%. External costs are hardly affected in this scenario, which primarily impacts direct production costs (see Figs. 7 and 8).

This has implications for policy and policy instruments. What may be effective on one route may be less so on another. Though the corridors analysed are notional routes, the segments and terminals within them are realistic segments for actual intermodal operations.

5.2.2. Rationalising Pre- and Post-Haulage Operations

Pre- and post-haulage operations are often considered as the weakest link of the intermodal chain. Although in distance terms they usually represent only a small fraction of the overall journey, they often penalise the overall door-to-door service, owing notably to their poor energy and environmental performances and their less than optimal load factor. This scenario has two components. The first deals with the effect of increasing the load factor on

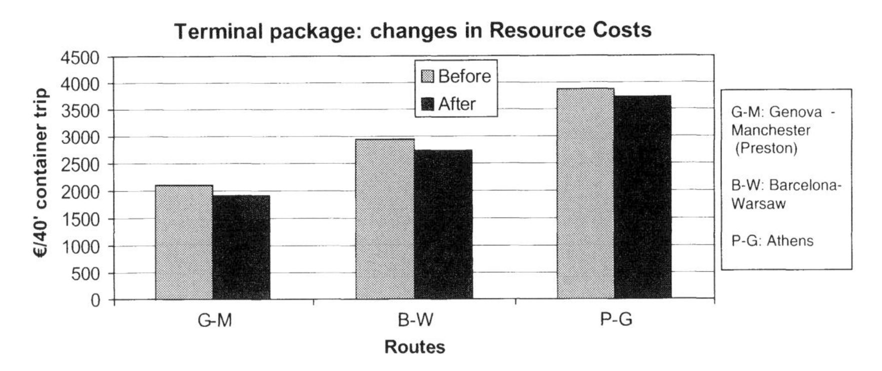

Fig. 7. Effect of Terminal and Transhipment Changes on Resource Costs per Trip.

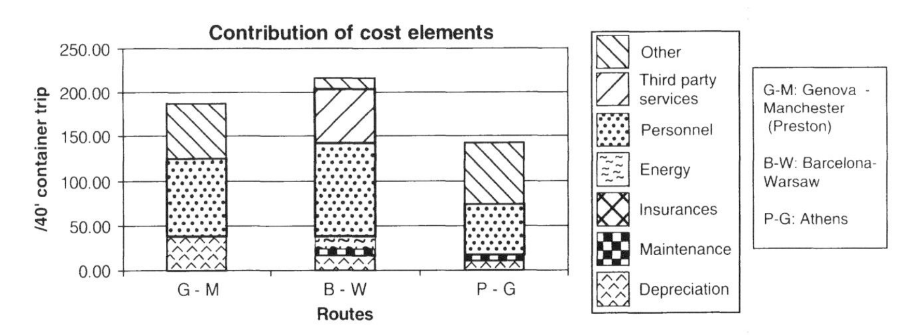

Fig. 8. Disaggregation of Reductions in Resource Costs.

pre- and post-haulage vehicles. The second analyses the scale of change in external costs if newer technology (Euro Class II) vehicles were systematically used in place of the older ones (Euro I or less).

5.2.3. Load Factor

The pre- and post-haul segment of an intermodal trip presents a difficult challenge. This stage often requires the use of heavy road haulage in urban areas. The demands for delivery to customers within certain delivery time windows and the lack of return loads lead to poor utilisation and poor load factors.

The critical factor required to reduce costs is the improved co-ordination of vehicle movements. In the short medium term, it can be assumed that the load factor for the road haulage collection and delivery vehicles can be increased from around 60–90% (typical also of long distance road haulage), notably through a massive recourse to information technology enabling real time access to operational data. This would then reduce the pre- or post-haul vehicle kilometres per 40-ft container kilometre from 1.67 to 1.11 with associated reductions in resource costs. The effect on the resource costs is shown in Fig. 9 which, in addition to the three RECORDIT corridors, presents the results for three shorter routes (Torino–Bologna, Milano–Munchen and Coventry–Rennes).

The overall effect in the three original corridors is an unweighted average of 2.2% reduction in resource costs door to door per 40-ft container trip and 4.1% in the three shorter routes. The reduction in taxes and charges associated with road haulage is in very similar proportions.

Although the effect of load factor on resource costs is somewhat limited, there are also benefits from the reduction in external costs as shown in Fig. 10. These vary from 3% for Genoa–Manchester route to 27.7% for Turin–Bologna with heavily urban and longer pre- and post-haul.

External costs are typically between 10% and 15% of the resource costs, so that the absolute value is not large but is indicative of the environmental impact of pre- and post-haul freight vehicles, especially in urban areas.

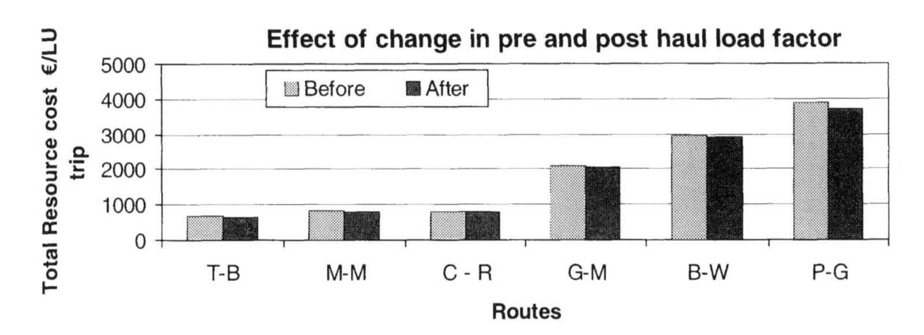

Fig. 9. Changing Pre- and Post-Haul Load Factor from 0.6 to 0.9.

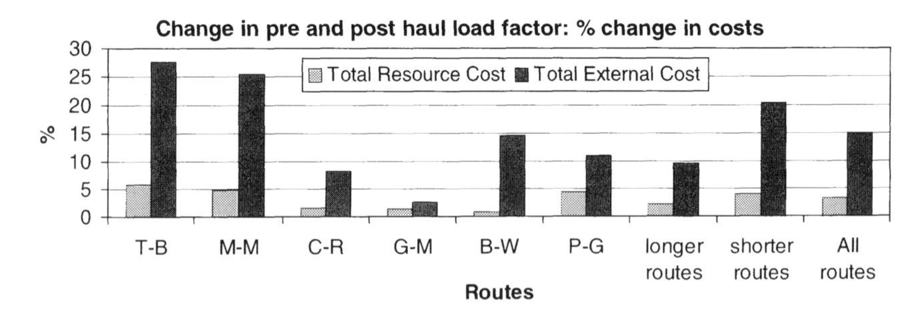

Fig. 10. Percentage Changes in Resource and External Costs.

The data also show the relatively larger effects on costs at shorter distances. The implication is that policies or actions need to be sensitive to origin-destination distance as well as to the other factors that differentiate routes.

5.2.4. Type of Truck Technology

As noted earlier, the pre- and post-haulage segment is not only a significant cost within the intermodal trip, but also takes place largely in urban areas. The vehicles used are therefore operating under the most difficult operating conditions and potentially are most polluting.

In this analysis, a simple substitution is made for current Euro I class trucks by Euro II class trucks in order to identify the reduction in external costs. The environmental effect is largely through a reduction in particulates in the exhaust. Fig. 11 shows the change in costs, in the range of €1–5 per trip.

It should be remembered that other activities in the intermodal transport chain also contribute to air pollution costs and that air pollution is only one of the external costs (typically and fairly consistently around 20%).

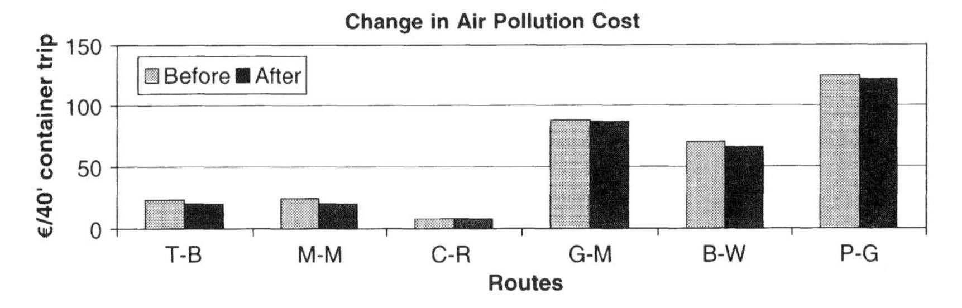

Fig. 11. Change in Air Pollution Cost Per 40-ft Container Trip.

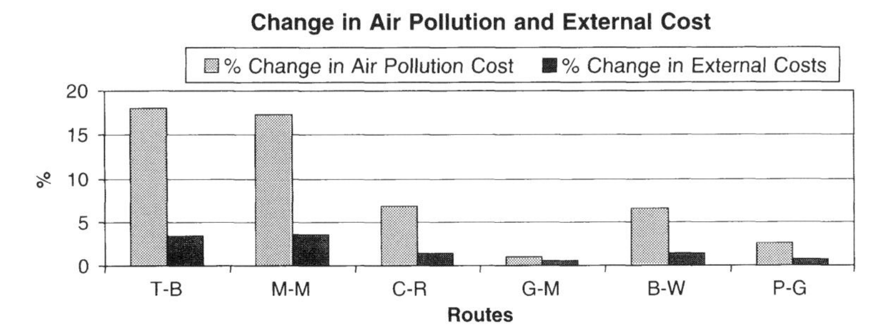

Fig. 12. Percentage Change in Air Pollution and External Costs.

Fig. 12 shows the proportional changes in air pollution and overall external costs.

Again the pattern between routes shows a marked difference between the shorter and longer routes as is expected. The pre- and post-haulage activity is a higher proportion of shorter journeys so that the shorter end of the intermodal market shows relatively bigger benefits from the new technology.

5.2.5. Rail Haul Improvements

The 2001 White Paper calls for a 50% gain in overall energy efficiency on railways by 2020. Most, but not all, services use electric locomotives. Translating this into possible actions at the technical and operational level gives the following package.

(1) *Increasing the number of wagons per train to 40 wagons.* This would reduce the energy consumed per LU kilometre though it would require substantial changes to layout of track in places. It would also provide more capacity per track slot. This is a medium- to long-term prospect involving marketing, and service design to increase demand and clever operational organisation.
(2) *Reduce starting and stopping.* There is a very significant energy input into starting a heavy freight train, yet they are currently subject to many stops and starts for traffic reasons. Increased dual track provision and prioritisation could reduce them ideally to zero. This long-term prospect partly involves capital expenditure, but also organisational efforts and, more widely, some prioritisation of freight trains.
(3) *Increase the load factor.* This assumes that the proportion of wagons loaded with revenue earning containers could increase to 95%. Since load factors are already 75–85%, this also depends on stimulating demand and clever scheduling. There are some underlying issues, not least that of the costs and price of returning empty containers where flows are unbalanced, which require further examination.
(4) *Energy efficiency of locomotives.* A reduction in kWh/train km through technical means (locomotive design, power distribution, etc.). This is across the range of time scales since some actions are already in hand and other medium- to long-term sequence are the subject of considerable EU and industry activity.

These four effects together lead to an average 16% reduction in resource costs. The magnitude of the resource cost savings is the source of potential price reductions (Fig. 13).

Fig. 14 shows each of the cost elements generated at a further level of disaggregation.

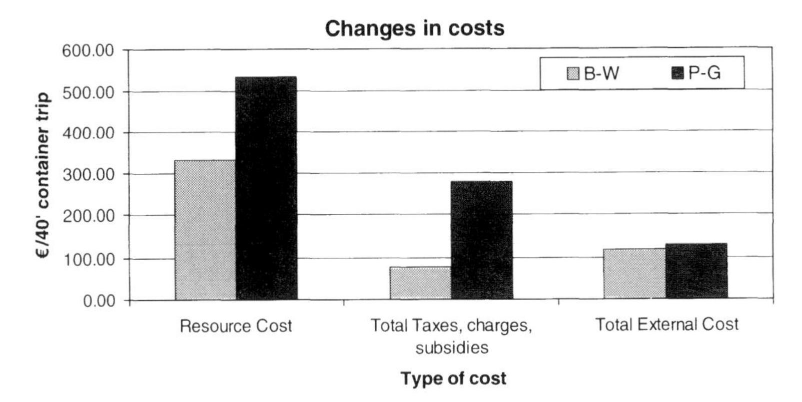

Fig. 13. Changes in Type of Cost by Route.

5.3. Internalising External Costs

In the perspective of the decisive pricing reform advocated by the 2001 White Paper and further discussed in subsequent EU policy documents (e.g. the Railway packages, the revision of the Eurovignette), the most obvious exploitation of marginal social cost data is the derivation of fair and efficient charges to reflect the application of the user pays principle through the internalisation of external costs.

A previous section has highlighted the current level of social cost coverage by existing taxation and charging regimes on selected corridors. Translating these results in terms of optimal charge to ensure the full coverage of Marginal Social Costs raises a number of methodological and practical difficulties.

5.3.1. Charging for Door-to-Door Intermodal Services

Ideal social charges can be calculated in €/container km, and, based on a set of assumptions on load factors and vehicle characteristics, these figures can then be expressed in €/tonne km, or in €/veh km. While unimodal transport services (road, rail) are directly amenable to the application of a charge expressed in one of these units, intermodal movements are currently carried out by a multitude of operators, including end-users (shippers and consignees), infrastructure managers, operators (modes, terminals) and service providers (forwarders, integrators), which interact at various levels and in non-standardised forms. Owing to the high number of operators involved in customer/client relationships, it is extremely difficult to establish a satisfactory (i.e. faithful, transparent and standardised) rule to avoid double

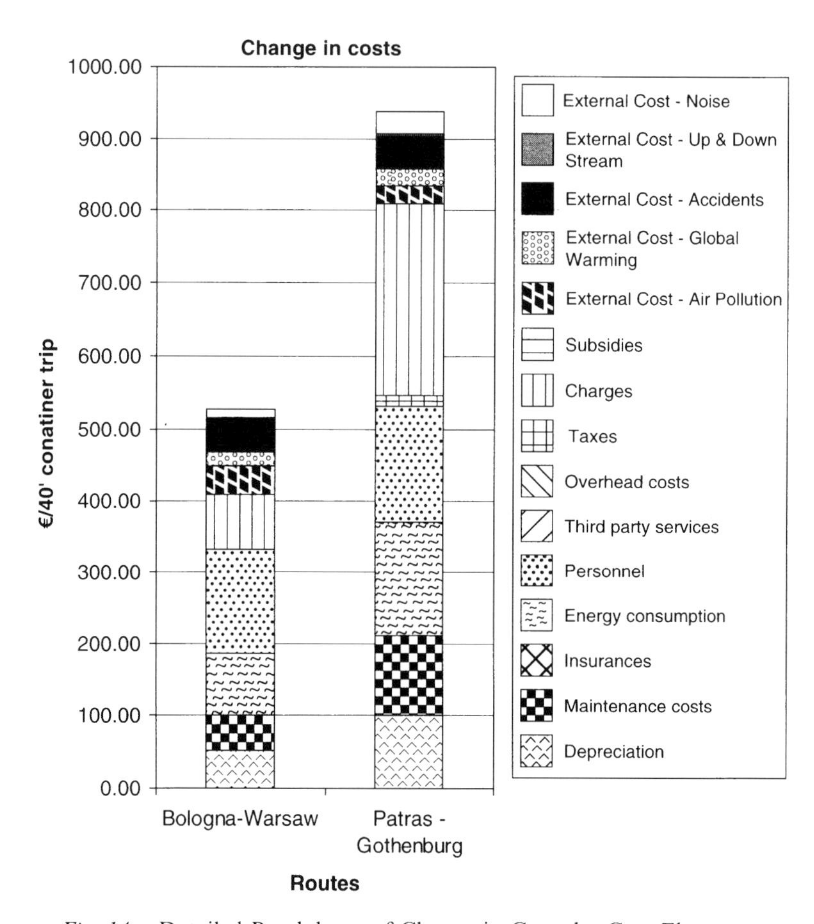

Fig. 14. Detailed Breakdown of Change in Costs by Cost Elements.

counting while ensuring the required completeness of information. The emerging role of the so-called integrators, while simplifying the picture by consolidating the process of service production, and thereby facilitating the application of a distance-based charge, may on the other hand induce an increased lack of transparency for the market users (whereby a significant part of transaction costs would become agency costs).

5.3.2. From Corridors to Countries

Previous sections have shown the derivation of taxation, external costs, infrastructure costs and payments for infrastructure for selected routes.

Taxation figures can be regarded as representative of the national taxation levels as they represent national taxation in each individual country.

Similarly, the infrastructure charges refer to averages in those countries. Payment for infrastructure is calculated from specific routes and may be a poor representation of average road tolls (or rail infrastructure charges) in the country as a whole.

Marginal external costs are also estimated for specific routes, and must be adjusted to account for the considerable variability encountered across different geographical, socio-economic and physical contexts:

(a) Air pollution and noise costs are known to vary considerably with time and location and, as a result, no direct extrapolation at country level is reasonable. Cost drivers such as e.g. vehicle technology, population density, average climatic conditions may however be identified to guide generalisation attempts.
(b) For what concerns congestion, marginal costs are usually calculated through a bottom-up approach, i.e. starting from the data collected at the level of each specific corridor and segment (length and duration of congestion episodes on that segment), which are then monetised through the application of values of time taken at the national level (what is in fact considered is the loss of time for other transport users that suffer from the congestion generated by the marginal (additional) truck). Extrapolation appears completely impossible here, as the frequency and gravity of congestion episodes is directly and exclusively related to the characteristics and conditions of the specific segment (with a high variability of the resulting cost values), and no pattern whatsoever can be established to relate one segment to the other.
(c) For global warming, the Impact Pathway Approach can be used, but the uncertainties that affect the bottom-up approach are too high (resulting in a range between minimum and maximum credible values that is as high as two orders of magnitude!). It is therefore common practice to adopt the so-called avoidance approach, as discussed in Chapter 7. In summary, the contribution of CO_2 to the total external costs of transport is valued the same wherever and whenever the transport activity takes place. The technology of the vehicle and, of course, the length of the corridor, are in fact the only parameters affecting the value of that cost component: no time- or location-dependency is considered, and, as a result, no generalisation problem arises.
(d) Wear and tear (maintenance) costs are also "top-down" values, i.e. derived from the overall costs of network maintenance at the level of each country.

These overall costs are then allocated to specific transport activities (e.g. the transportation of a given LU, on a given road, with a given truck), through statistical techniques (see Chapter 2). The result is a table of values for each country, giving a set of estimates differentiated according to the above-mentioned parameters. Although the accuracy of these values can indeed be questioned, adopting this method is often deemed the only practicable solution, considering that the alternative would require the gathering of micro-data at the level of each network segment describing the specific maintenance costs that such segment requires. With such an approach, no generalisation problems arise for wear and tear.

Based on the above guidelines to country level generalisation, the following Table 5 shows a possible derivation of charges for Heavy Goods Vehicles (HGV) (€/veh km, 2001 values).

5.4. Impacts of Internalisation on Intermodal Demand

The benefits of marginal cost pricing from an efficiency perspective depend critically on the response of decision makers to any change in prices. If the response to tax increases is limited in terms of modal transfer and reducing

Table 5. Extra Charge for HGVs to Cover MSC.

Country	Taxes	External Costs	Infrastructure Net Payment	Extra Charge
Austria	0.14	0.36	−0.13	0.35
Croatia	0.12	0.26	−0.06	0.19
Denmark	0.09	0.33	−0.09	0.33
France	0.14	0.31	0.00	0.17
Germany	0.13	0.30	−0.05	0.22
Greece	0.17	0.40	−0.07	0.30
Hungary	0.11	0.35	−0.06	0.29
Italy	0.09	0.30	−0.01	0.22
Netherlands	0.15	0.29	−0.08	0.22
Poland	0.14	0.28	−0.06	0.20
Slovakia	0.12	0.39	−0.06	0.33
Slovenia	0.12	0.54	−0.05	0.47
Spain	0.12	0.33	0.00	0.21
Sweden	0.09	0.24	−0.04	0.19
Switzerland	0.16	0.36	−0.15	0.35
UK	0.34	0.36	−0.18	0.20
Average	0.15	0.32	−0.05	0.21

overall movement, then the benefit will be small. The benefit, therefore, depends on the elasticity of demand.

If it is assumed that the price of road freight transport increases with other prices remaining unchanged (due to a fuel tax increase for instance), then there are two distinct impacts. In the first place, the demand for movement by competing modes should increase. And secondly, the total demand for freight (measured in tonne km) should fall. This latter effect might be expected to take rather longer to materialise. The main source of any reduction in total movement can be expected to stem from a fall in average haul. This is likely to be far more important than changes in tonnes lifted (which is intimately connected to the level of output in the economy) or better utilisation of vehicles (this has been remarkably stable over a long period without any apparent change due to previous changes in real costs). One estimate (Baum, Wells, & Lund, 1990) of the elasticity of road freight with respect to price suggests a range between −0.2 and −0.7 for different commodities with a mean near to −0.6. These figures refer to trips greater than 50 km. Following this study, there has been only limited work undertaken to refine these estimates.

Turning to the second impact – a transfer to other modes – this was also examined in the same report (Baum et al., 1990). Cross elasticities (the percentage change in rail demand with respect to the price of road) revealed a wide range for commodities from 0.2 to 1.9. The (unweighted) mean was near to 1.0. A number of studies carried out in Europe over the last 10 years have developed more complex modal split models of the freight market. Rather than time series analysis of freight demand, they have concentrated on cross section analysis usually focussing on a set of specific routes.

The study offering the greatest insight into the intermodal market in Europe was carried out by Lobé (2001) in 1999–2000. Examining 12 routes in Europe, the study found that although there are many factors influencing demand, price is still critical. Derivation of an elasticity is complex. It varies for different routes due to the different distances and different modal shares. Any assumption about an average elasticity for intermodal transport in Europe is therefore disguising a wide variation on different routes. Using the output from Lobé it is possible to deduce that an average cross elasticity (intermodal demand with respect to road price) might be of the order of 1.7. Elasticity estimates for Inland Water and Short Sea Shipping based on the same study are 1.2 and 0.5, though the small sample size means the estimates are subject to a wide margin of error.

Some support for the intermodal cross elasticity is found in Beuthe (2001). This suggests that the relevant cross elasticity for rail (rather than intermodal) is around 2. There seems no reason to believe that the intermodal component of rail transport would not experience these elasticities also. This model, along with the others, also assumes that an increase in the price of road movement will only lead to a transfer to existing intermodal services. In practice it is possible that the intermodal industry will respond to the opportunities offered by higher road charges by introducing new services, thus attracting extra demand. Taking account of these various studies for the calculations that follow a cross elasticity of 1.7 and an own elasticity of −0.3 in response to a change in the price of road freight are adopted.

The first set of calculations are made under the assumption that the target is to increase the tax paid for road operations by the equivalent of €0.17/container km using a simple fuel tax increase. In the intermodal market of interest here, using articulated road vehicles with a utilisation factor of 0.85, this would require an increase in fuel taxation of approximately €0.44/l, which represents more than a doubling of the original tax rates existing in 2000. Any such increase in taxation will increase the cost of both road and intermodal solutions (due to the share of pre- and post-haul). Assuming that the base prices in the market are represented by internal costs (€1.05/40′ container km for intermodal), then the increase in demand for intermodal transport would be about 20%. Using the 1997 figure of 57.8 billion tonne km (Btkm) by rail intermodal, this represents nearly 12 Btkm extra. This can be compared with the total figure of 60 Btkm that the White Paper projects will be transferred to rail as a result of policy initiatives (European Commission, 2001).

The benefits of the transfer of freight from road to rail is a saving in resource costs, consignments are transferred from road with its high external costs to intermodal with its lower external costs. The welfare benefit to members of the EU of the modal transfer in money terms can be calculated as €60 M per year (the calculation is based on the welfare concept underlying the demand curves for intermodal and road transport).

Significant benefits also flow from the reduction in movement on the road associated with a reduction in the average haul. The response to price is low (elasticity −0.3) in this case, which leads to benefits of about €40 M per year (this refers to road movements of unitised transport only).

6. CONCLUSION

This chapter has addressed the specific task of estimating the marginal social cost of road and of intermodal transport. This is not an easy task, given the large number of component costs, the variability of costs with location and vehicle type and the uncertainty regarding the valuation of external costs. However, it is achieved for three long corridors covering a wide variety of countries and circumstances. Despite wide variability, it is typically found that full internalisation of external costs will benefit intermodal transport, as would various other measures to raise its efficiency.

REFERENCES

Baum, H., Wells, J. R., & Lund, A. K. (1990). *Aufbereitung von Preiselastizitaten der Nachfrage im Guterverkehr fur Modal Split Prognosen*. Essen: Untersuchung im auftrag des Verkehrsforums Bahn e.V.

Beuthe, M. (2001). *Freight transport demand elasticities for forecasting modal shifts*. THINK-UP Seminar in Berlin.

European Commission. (1999). *Final Report*. REDEFINE.

European Commission. (2000). *EU Transport in figures*. DG TREN.

European Commission. (2000). *Report 1*, SOFTICE.

European Commission. (2001). *European transport policy for 2001: Time to decide*. COM 2001 (370), Brussels.

Link, H., Dodgson, J. S., Maibach, M., & Herry, M. (1999). *The costs of road infrastructure and congestion in Europe*. Heidelberg: Springer.

Lobé, P. (2001). *Computation of a perception index in intermodal transport*, PTRC.

RECORDIT www.recordit.org

MEASURING MARGINAL SOCIAL COST: METHODS, TRANSFERABILITY ☆

Marten van den Bossche, Corina Certan, Simme Veldman, Chris Nash and Bryan Matthews

1. INTRODUCTION

In this volume, we have outlined the methods available for best practice estimation of marginal social cost. The UNITE (UNIfication of accounts and marginal costs for Transport Efficiency) project, which generally represents the state of the art on this issue, involved identification and implementation of these methods by means of case studies including provision of new empirical evidence, and building on existing empirical evidence.

However, it cannot be expected that the data, expertise and time will always be available to apply these best practice methods when pricing decisions have to be made, hence, there tend to be only a small number of studies that are fully up to date and reliable and some means of generalising or transferring findings is essential. The UNITE project, therefore, also involved examination of the possibilities of generalisation and transferability of methodology and

☆ This chapter in particular owes much to the contribution of other UNITE partners: in particular John Nellthorp and Dan Johnson (ITS), Andrea Ricci and Riccardo Enei (ISIS), Peter Bickel and Rainer Friedrich (University of Stuttgart) and Gunnar Lindberg (VTI).

Measuring the Marginal Social Cost of Transport
Research in Transportation Economics, Volume 14, 287–314
Copyright © 2005 by Elsevier Ltd.
All rights of reproduction in any form reserved
ISSN: 0739-8859/doi:10.1016/S0739-8859(05)14010-4

results to other contexts, and elaboration of guidelines for the generalisation of the marginal costs estimates.

In considering whether and to what extent results of marginal cost case studies can be generalised to other situations, the UNITE project sought to examine five key questions:

- How consistent are estimates for the same cost category across different situations?
- How can estimates in a specific case be adjusted to account for different circumstances?
- What is the relationship between the cost driver and the respective estimate?
- What technical and organisational issues pose problems for generalisation?
- How practical and theoretically robust are the transferability approaches proposed?

In this penultimate chapter, we give an overview of empirical results, briefly review the conclusions from the work on measures of marginal social cost and the potential for transferability and generalisation. Conclusions for policy are then explored in the final chapter.

2. EMPIRICAL RESULTS

In this section, we present an overview of some of the empirical results from the UNITE case studies, and seek conclusions on the relative importance of the different cost elements by mode and context as well as the degree to which the magnitude of estimates for that cost category varies. It should be noted that in general each cost category for each mode was studied in a different case study, so that the various cost categories cannot simply be added together. Moreover, the output variables that were used in the different studies varied, so to make them comparable various assumptions had to be made. The bar charts should therefore be seen as illustrative of general tendencies rather than as precise numerical estimates. Where a range of results is produced by different case studies, bar charts show the upper and lower values.

Figs. 1a and b summarise the results for car travel. It is clear that for car the dominant element is congestion (especially for urban trips), however, this varies greatly between case studies. Accident costs are also comparatively large, particularly in urban areas. Also, in urban areas noise is very

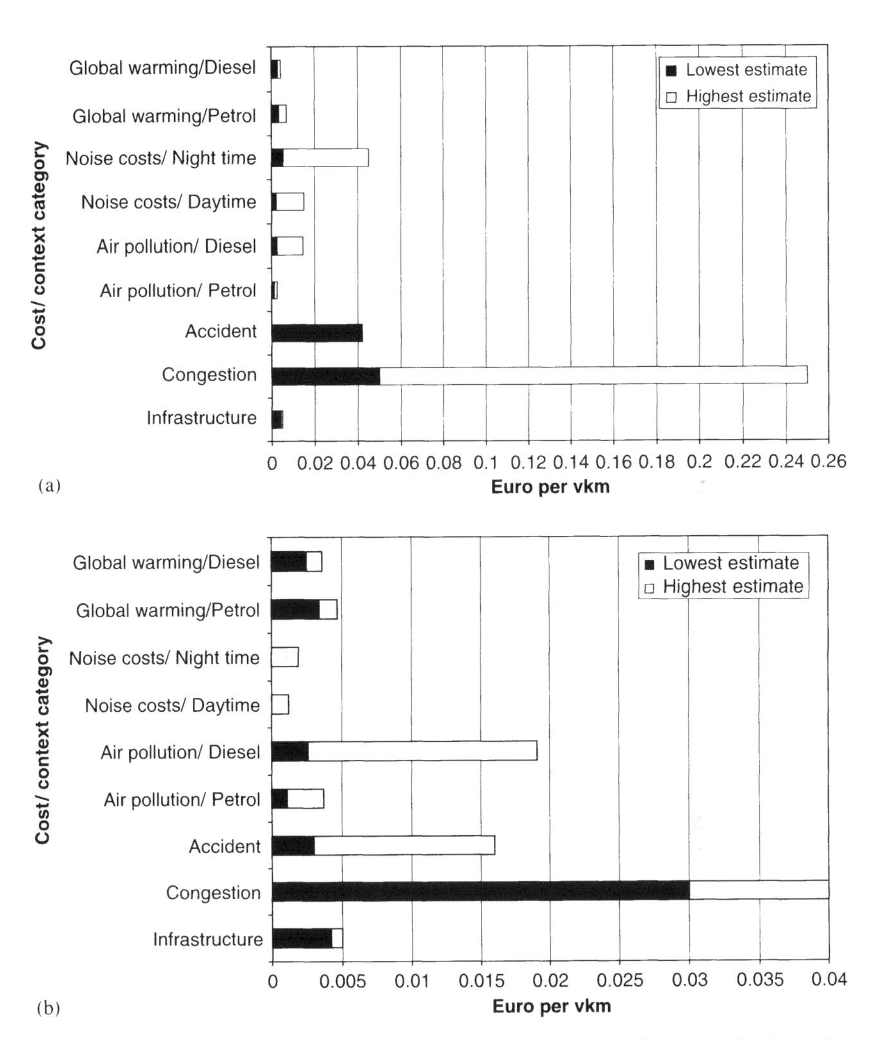

Fig. 1. (a) Overview of MC for Urban Car Travel (Euro per vkm) and (b) Overview of MC for Interurban Car Travel (Euro per vkm).

important, particularly at night, and air pollution is also significant. Many of the environmental costs also vary due to differences in traffic densities and speeds. Areas of high traffic speed and density report lower marginal noise and air costs, as an additional car has less impact than in quieter areas.

Marginal infrastructure costs, air pollution and global warming costs for car are all relatively small.

The most striking feature of these diagrams is the enormous variation from case to case for the costs of congestion, noise and air pollution. A point we return to when considering transferability and generalisation issues below.

Figs. 2a and b produce the same diagrams for heavy goods vehicles. Here congestion costs are similar as for car, but a smaller component of the overall marginal cost, while infrastructure costs are somewhat greater. Noise costs are important for urban Heavy Goods Vehicle (HGV) travel, and air pollution costs remain important for inter-urban routes, and in most cases these are over a factor of 10 higher than for cars. One urban study in a low-density traffic area finds night-time noise the dominant factor.

Figs. 3a and b show the results for rail passengers. Naturally, supplier operating costs are the dominant elements, and are many times higher in the peak than in the off peak. Urban congestion and inter-urban air pollution costs are the next most significant items. Air pollution and global warming costs varies with type of traction and (where relevant) prime energy source for electric traction. Marginal infrastructure costs for passenger services are low. It should be noted that in the inter-urban rail case study, supplier operating costs are strictly for lengthening trains; for purposes of comparability they have been expressed per train kilometre given typical train loadings but this is somewhat misleading given the nature of the case study. Also when capacity is expanded solely by lengthening trains there is no Mohring effect, so these two cost categories certainly cannot be added together.

Fig. 4 shows comparable figures for rail freight, except that we do not have a supplier operating cost case study for freight. Air pollution, global warming and noise are all of similar orders of magnitude to marginal infrastructure costs. Infrastructure costs are higher than for passenger rail due to the higher gross tonne kilometre per train of freight.

Finally, Fig. 5 shows the results for air transport. Of course supplier operating cost is dominant, but these results suggest that both congestion costs and the Mohring effects are more important than environmental effects. In circumstances where line density is low, the Mohring effect is quite important, leading at least to an a priori case for subsidy for scheduled air services. (For consistency across cost categories, we have assumed an average journey length of 930 km as was used in the environmental cost case study.)

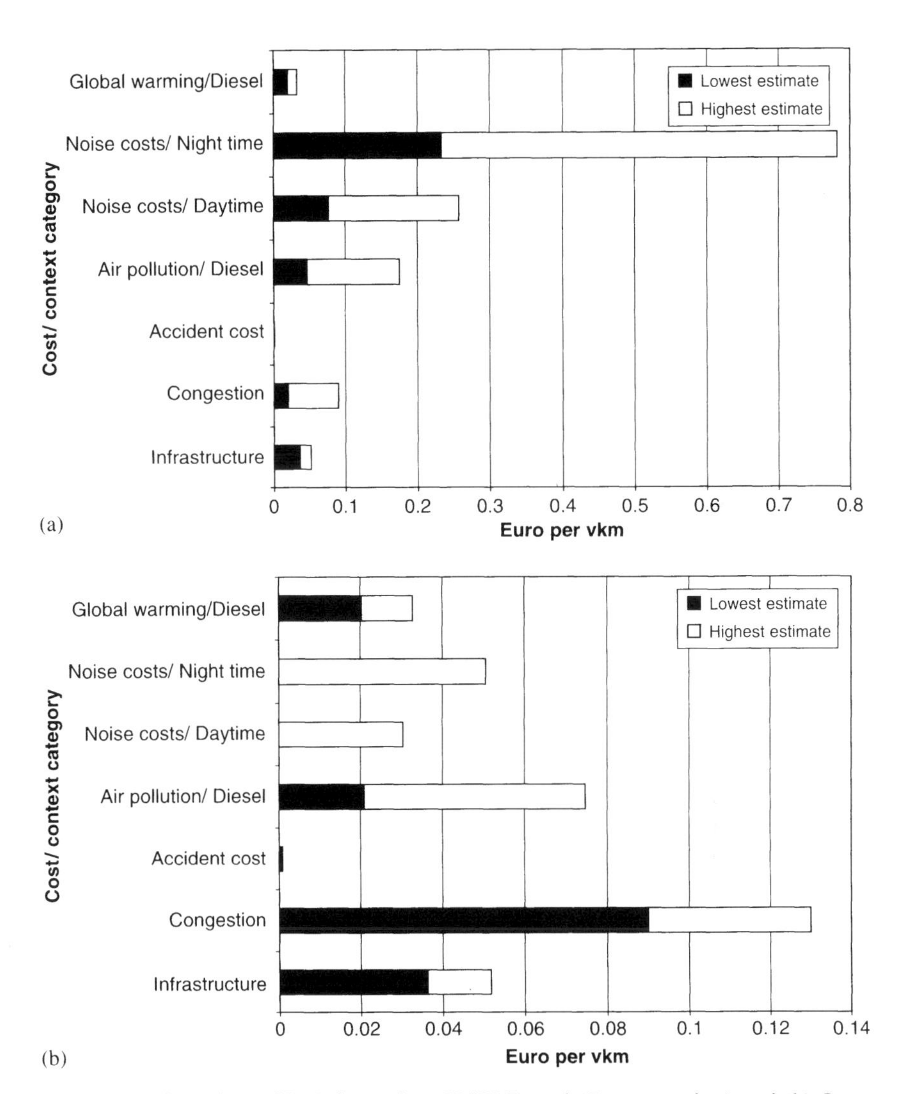

Fig. 2. (a) Overview of MC for Urban HGV Travel (Euro per vkm) and (b) Overview of MC for Interurban HGV Travel (Euro per vkm).

3. MEASUREMENT AND TRANSFERABILITY

In general the estimation of marginal cost may be broken down into estimation of a physical effect, and its monetary valuation as follows

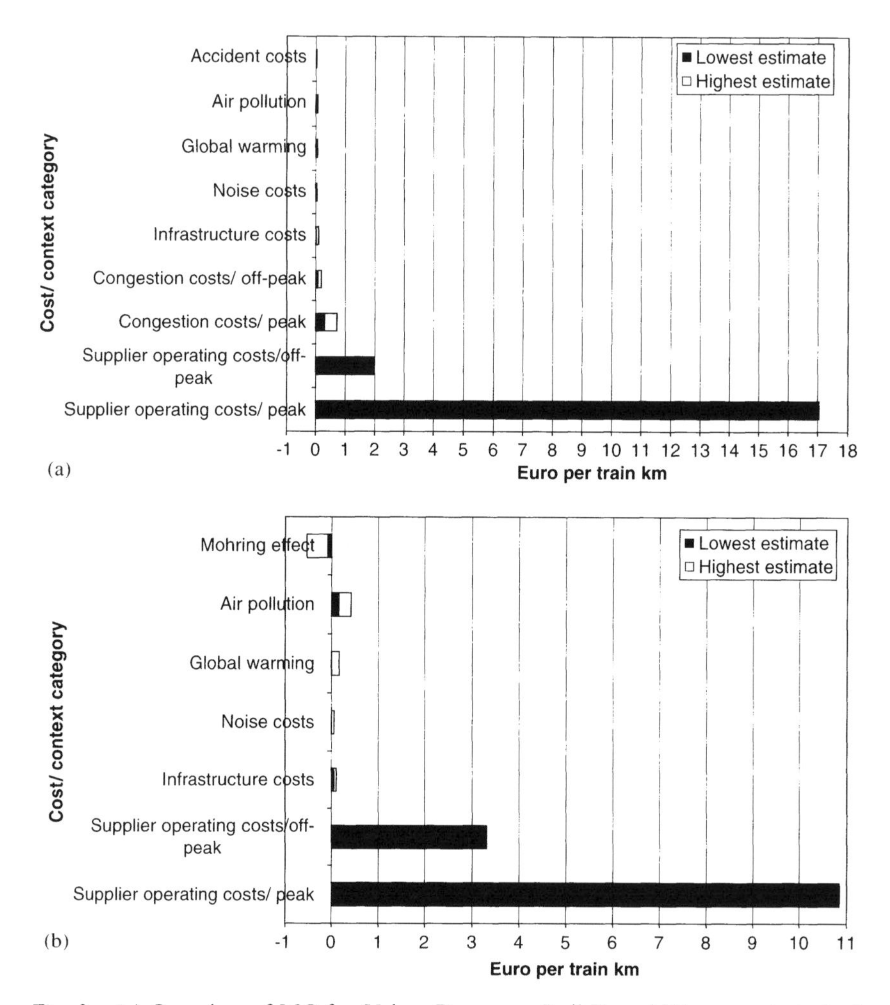

Fig. 3. (a) Overview of MC for Urban Passenger Rail Travel (Euro per train km) and (b) Overview of MC for Interurban Rail Travel (Euro per train km).

(sometimes the two steps are combined in a single equation):

$$\text{marginal cost} = \text{physical effect} \times \text{monetary valuation.}$$

The monetary valuation of a given physical effect will in general vary between contexts for a relatively small number of reasons – different price

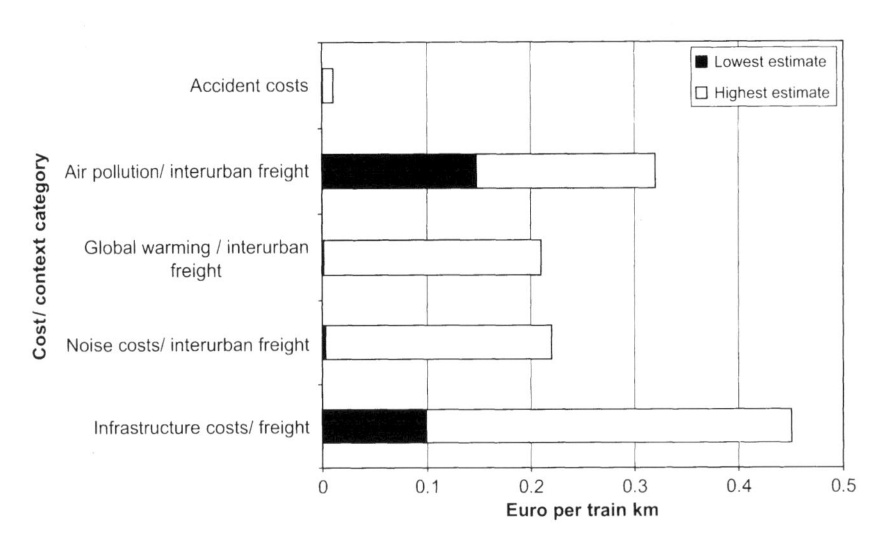

Fig. 4. Overview of MC for Rail Freight Travel (Euro per train km).

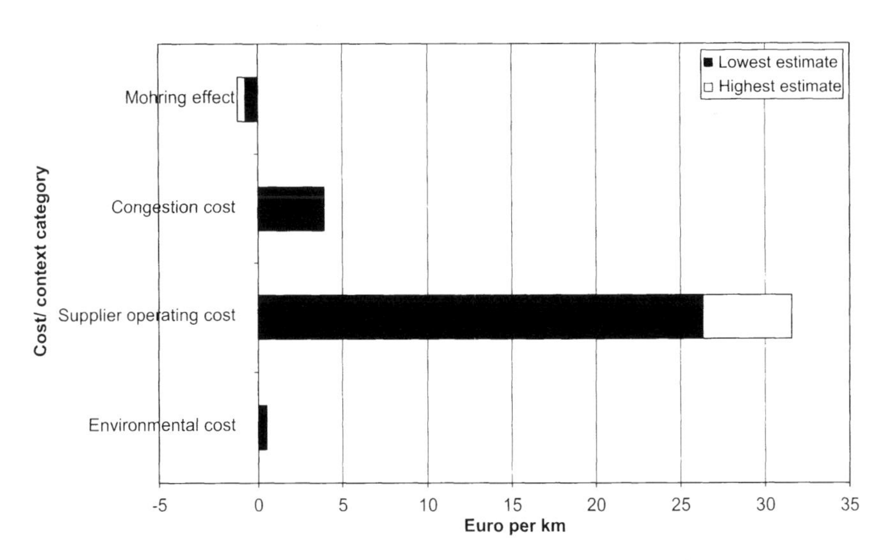

Fig. 5. Overview of MC for Air Travel (Euro per vkm).

levels, different relative prices, different real incomes and different tastes. Where there is confidence that the only difference in context is one leading to different money values, techniques for translation of the money value from one context to another can be used to transfer the entire result from one context to another. But this will be exceptional.

In most cases the physical effects will differ. Reasons for such differences include:

- Different starting points in terms of the current situation (nature of assets, traffic volumes, background conditions, etc.).
- Different population, population density and locational patterns.
- Different climate.
- Different laws, customs or behavioural patterns.

In these situations, generalisation may relate more or less to transferability of methodology, transferability of key functional relationships, transferability of input values to these relationships or transferability of actual output values. Clear guidance on all of these aspects of transferability and generalisation is keenly sought by policy-makers and practitioners to enable them to use appropriate techniques for transferring state-of-the-art methodologies, functional relationships, input values and output values to the characteristics of their particular context.

In general we would expect methodology to be transferable from one location to another. The main limits on this are the availability of data, special circumstances that may need to be taken into account in any particular example and the sheer cost of applying best practice methodology for each and every calculation needed. Where data availability or cost rule out the use of best practice methodology in each case then consideration needs to be given to a simpler alternative methodology which may then be used – perhaps with information such as cost elasticities from the best practice examples.

If a functional relationship, such as a speed-flow curve, complete with its parameter values, may be transferred from one context to another then information relating to the new context (in this example traffic flow, number of lanes) may be used to calculate the value relevant in the new context. Where the parameter values differ but the functional form is believed to be transferable, then a way must be found to recalibrate the relationship in order to transfer it.

In general, the input values to the relationship (e.g. in the above example traffic flow, number of lanes) are the factors most likely to be context specific, and also the factors for which context specific information is most readily available. However, there may be some circumstances in which input data may be usefully transferred between contexts.

Also, here the aim should not solely be a description of the type of function to be used, but the objective should be extended to an assessment of how specific functions derived from empirical research directly can be adapted to a new specific context. If a cost function is produced, how can this be (re-)used in an appropriate form?

Most attractive for direct use are, of course, values found in empirical research. Transferability in this case means that a value in effect can be adapted to a different context. In particular values refer to: basic inputs to the methodology (e.g. vehicle emissions), economic unit values (e.g. the value of life, the unit cost of construction), output values (that is empirical outputs), output ratios or relationships (for example the ratio of marginal to average cost).

Where circumstances are comparable, actual outputs may be transferred from one context to another, with adjustment confined to differing economic valuation as discussed above. If actual outputs are not directly transferable, it may be possible to transfer by applying some adjustment, allowing for factors such as traffic or population density.

In the cost categories specific sections we consider the possibilities for generalisation in terms of the above three levels for each of the cost categories in turn. We then consider transferability of money values in the subsequent section.

3.1. Generalisation of Marginal costs of Infrastructure Use

As was seen in Chapter 3, we have two preferred methods for estimating the marginal cost of infrastructure use: econometrics and engineering. The problem with both is getting suitable data at a sufficient degree of disaggregation.

The ideal approach is to transfer the econometric methodology and to regress the relevant cost elements on traffic levels and other relevant variables using data for the application at hand. However, data requirements for this are severe. It requires data on maintenance and renewal costs for individual segments of infrastructure, and corresponding data on traffic levels disaggregated by vehicle type. These data are often not available; even if experience suggests that multicollinearity may make it impossible to get useable results. It will certainly be impossible to get detailed results for a large number of different types of vehicle (e.g. different axle weights) in this way. While engineering studies may shed light on these types of issues (e.g. the relative damage done by different types of vehicle) it is relatively difficult to build up a full picture of costs in this way.

It may be therefore that both econometric and engineering approaches have to be seen as informing a more traditional cost allocation approach, rather than as approaches to be routinely used in all cases. However, in many countries infrastructure management reforms have resulted in data being recorded at a more disaggregated level, making it more amenable to the application of econometric techniques.

The actual value of marginal cost will clearly differ from location to location according to factors such as variations in factor prices, transport costs, weather and ground conditions. Lindberg (2002) reported that, for road and rail, the cost structures follow a parallel pattern, with 'low standard networks' being characterised by higher marginal wear and tear costs and 'high standard networks' by lower marginal wear and tear costs; moreover marginal costs are generally below average cost. A number of national studies exist on rail track costs, suggesting that the similarities between countries are considerable but not systematic. Few published studies are available on airfield wear-and-tear though and we cannot judge on the magnitude and structure of this cost component. For inland waterways and maritime transport the infrastructure maintenance cost is in general small.

In terms then of transferring results from one context to another, it seems more plausible that the cost elasticity, being a ratio of marginal to average cost, may be transferable. With the cost elasticity, combined with data on average costs which is, in general, fairly readily computable, marginal cost may be estimated relatively straight forwardly. However, confidence in transferring cost elasticity values requires there to be some consistency among cost elasticity estimates for different contexts.

For roads, the evidence on this parameter is mixed. The German case studies showed elasticities of somewhat less than one (for maintenance and renewals of the order of 0.8), while the Swedish engineering cost approach had an average elasticity for pavement reconstruction and resurfacing as low as 0.4 (but varying from 0.1 for high-strength roads to 0.8 for low strength). The differences may be because the elasticity rises with traffic level.

For rail, there was greater consistency, with an elasticity of maintenance and renewals costs with respect to traffic levels of the order of 0.14–0.17 emerging from the econometric analysis and 0.2–0.3 emerging from the engineering approach. The higher value for the latter may be explained by the higher levels of traffic and speed in Great Britain (to which the case study relates). It appears from the review of international experience conducted for the British rail regulator that the elasticity may be higher where traffic volumes and speeds are high. A new econometric study for France (Gaudry & Quinet, 2003) gives elasticities in the range of 0.3–0.4. The cost elasticity will be lower where structures, signalling and electrification account for a higher proportion of total maintenance and renewals costs. Relative costs of different vehicle types will again require an engineering cost approach, with the British results from the Railtrack model perhaps providing some guidance.

For other modes we have little firm evidence, but it appears that the marginal costs of using existing infrastructure for air and water transport are generally very low.

If we do apply elasticities to average cost information to obtain our estimates, then of course information from accounts becomes a prime source of information. However, it is desirable that the accounts information should be as disaggregate as possible both in terms of type of asset (e.g. type of road) and in terms of categories of cost (e.g. track renewal, bridge maintenance).

3.2. Generalisation Aspects of Supplier Operating Costs

We saw in Chapter 4 that researchers tend to use an econometric approach for estimating supplier operating costs, while the industry itself uses cost allocation. Econometric approaches have the advantage that they easily deal with economies of scale and density, but the disadvantage that the level of outputs often is too aggregate to deal with issues relevant to pricing such as peak and off-peak differentials, and that the costs are difficult to obtain from accounting data. The cost allocation method is an approach which makes assumptions as to which categories vary with which output measures. However, it does not deal adequately with the effects of peaks on staffing levels; a better approach would be to do a complete vehicle and crew scheduling exercise to identify staff requirements and costs.

Most econometric results from studies of supplier operating costs tend to suggest roughly constant returns to scale with respect to firm size, but major economies of density of traffic, mainly arising because of economies of scale in the size of vehicle or train. The approximately constant returns to scale and comparable results suggest that the cost allocation method may be adequate. The sort of model this leads to is shown in the following equation:

$$\textit{Total Cost} = a + b * \text{train-km} + c * \text{train-hours} + d * \text{vkm} + e * \text{vehicle-hours} + f * \text{peak-vehicle requirement}$$

Careful thought is needed, however, as to how to use this information in marginal cost estimates. For instance, when additional traffic arises, in general the cheapest approach for the operator to handle this traffic is by increasing vehicle size or train length. Where this is not possible (for instance, because of infrastructure constraints that are too expensive to overcome), then service frequency is increased instead, giving rise to the so-called Mohring effect, whereby additional traffic benefits existing users by improving service levels (this may also arise because it is more profitable or socially desirable to increase service levels instead of vehicle size). Ideally, a complete scheduling exercise needs to be undertaken to identify the optimal service level, but simpler approximations may have to suffice. Jansson, in Chapter 5 of this book, shows both from theoretical and empirical evidence that suppliers will tend to increase service frequencies in proportion to the square root of traffic. However, he argues that, if the supplier is optimising service levels, and if increases in vehicle or train size are feasible, the marginal social cost will be the same however the supplier reacts. In this event we can simply compute the marginal costs of increasing vehicle or train size (in practice, because of indivisibilities it is frequently necessary to estimate the marginal cost per passenger kilometre as the average incremental cost of additional capacity. In that case, estimating an appropriate load factor, recognising that even in the peak, vehicles may only fill up on route, and may need to make a counter peak journey as well, is a crucial step. If, however, these assumptions do not hold, then it is necessary to consider how the supplier will react in practice, and to compute both the supplier operating cost and the Mohring effect for the change.

3.3. Generalisation Aspects of User Costs and Benefits

For road congestion the usual approach involves using speed/flow and junction-delay relationships, either in a model of specific links or in a complete network model, to model traveller behaviour so as to capture the range of responses made in the fact of congested conditions. These speed/flow and junction-delay relationships have been thoroughly researched and it is not usually necessary to undertake new research; many existing models are available which may be used. There is considerable variability in the estimates relating to different places generated by this approach though, and concerns arise that the variability might be linked to the type of model used – its degree of aggregations, its definition of the network etc. – rather than linked to actual differences between congestion in those different places.

Another source of variability may arise out of simple differences in the ways in which geographical entities such as the city centre is defined from one place to another or from one study to another. There are also the usual concerns about data availability required for such modelling exercises.

Other modes in general are much more difficult. There is little research to go on. The conditions in each rail route, station, airport, seaport or inland waterway port are different, as is the mix of traffic. In principle, for each of those facilities a characteristic relationship of traffic volume and delay must be estimated, taking account of the details of layout and traffic mix.

There is currently very little work available on the measurement of scarcity costs, which arise in scheduled modes where the allocation of slots to one user may mean that other users cannot get the slots they wish for, and none has been reported in this volume. The extent to which pricing reform can contribute to the efficient allocation of scarce capacity in rail and air remains uncertain because of the complexity of cost measurement and implementation in pricing policies. The most attractive solution to this problem of scarcity is, in theory, to 'auction' scarce slots, but many practical difficulties exist. Scarcity costs might be calculated directly. For instance, if a train or plane has to be run at a different time from that desired, it is possible to use studies of the value people place on departure time shifts to estimate the value to customers of the cost involved. Another possibility is to simply impose a price and see what happens to demand, and then iterate until demand equals capacity. The risk is, however, that serious distortions may occur while the price is adjusting, and that strategic game playing may occur to force the price down by withholding demand, where competition is not strong. Some countries, including Britain allow for a degree of 'secondary trading' in which slots change hands between operators at enhanced prices, but there are fears that this may pose competition problems. Given the difficulties with all these approaches, it may be that the best way of handling the issue is to permit direct negotiation between operators and the infrastructure manager over the price and allocation of slots, including investment in new or upgraded capacity. However, experience of this approach is that it is complex and time consuming given the number of parties involved and the scope for free-riding. It is also difficult to ensure that this does not lead to the abuse of monopoly power, particularly when the infrastructure manager and the operator are part of the same company.

We saw, also in Chapter 5, that measurement of user benefits arising from the 'Mohring effect' depends on the extent to which operators increase services in response to increasing traffic. If they do not increase services, the result is simply increased load factors, then there is no Mohring effect and the marginal cost to operators is close to zero (although there may be a disbenefit to passengers from increased crowding to take into account). If operators maintain load factors by operating larger vehicles then there is no disbenefit to passengers but there is a significant marginal operating cost, which, however, is typically well below the average cost. Jansson, in the case study on Swedish Rail finds theoretical reasons for expecting frequency of service to increase in proportion to the square root of the level of traffic, and shows that this indeed appears to be broadly the policy followed by Swedish Railways.

If we make the assumption that operators increase service frequencies in direct proportion to increases in patronage, a basic approach to estimating the Mohring effect is set out in Sansom, Nash, Mackie, Shires, and Watkiss (2001). The steps in the calculation begin by estimating the marginal external benefit of increased passengers as the number of existing passengers (Q) multiplied by the change in their average cost (AC) with a change in passengers:

$$MC = Q\frac{\partial AC_{\text{user}}}{\partial Q}$$

An increase in frequency allows the user to choose a more favourable departure time. We can rearrange this in terms of headway time, the time span between consecutive trips for an origin – destination pair. The mean value of this, assuming evenly distributed desired departure times and that users are equally willing to adjust to earlier and later departures, can be taken as quarter the headway (h), multiplied by the value of time for departure time shifts ($VOT_{\text{dep\_time}}$). The above equation then becomes

$$MC = Q\frac{\partial}{\partial Q}\left[0.25hVOT_{\text{dep\_time}}\right]$$

Furthermore, if occupancy (O) is fixed and Q is expressed in terms of passengers per hour, the headway equals O/Q and the equation becomes

$$MC = -0.25\frac{O}{Q}VOT_{\text{dep\_time}} = -0.25hVOT_{\text{dep\_time}}$$

The negative sign indicates that this is a marginal external benefit rather than a cost. This is the basic and very simple equation which may be used to estimate the Mohring effect in these circumstances. It should be noted that this is obviously a gross simplification of reality – in practice frequencies will not vary continuously with increases in traffic, but in discrete steps. However, if we are estimating the overall marginal social cost of additional traffic this is less of a problem than it appears; when frequencies do not increase we will have overestimated the Mohring effect, but we will have overestimated the marginal operating cost as well, and the sum of the two should be more stable. Moreover, Jansson shows that if frequencies are optimal then the marginal social cost should be the same, whatever assumptions we make about the way in which frequency adapts to traffic levels.

The results of Wardman (1998) suggest that the mean value of departure time shifts may be taken as 0.72 of the value of in-vehicle time. If service frequency is equal to or less than every 10 min, however, passengers are likely to turn up at random so that the departure time shift becomes wait time and, again according to Wardman (1998) this should be valued at a multiple of 1.6 of the value of in-vehicle time.

If frequency rose in proportion to the square root of Q rather than Q, then it is easily shown that the Mohring effect is halved (but so too would be the marginal operating cost, as half of any increased traffic would be accommodated simply by raising load factors). Note that the above formulation only applies to low-frequency services where passengers use a timetable. Where passengers arrive randomly the familiar half the headway formula should be used together with the value of waiting time rather than departure time shift.

3.4. Generalisation Aspects on Marginal Costs of Accidents

The risk elasticity method and theory presented in Chapter 6 is summarised in the equation for marginal external accident cost below:

$$MC_j^e = r(a + b + c)[(1 - \theta) + E] + \theta rc$$

where r represents accident risk, a the value of statistical life (VOSL), b ditto for relatives and friends, c the costs for the rest of society, θ the injurer's risk and E the risk elasticity (i.e. the relationship between accidents and traffic volume).

It is clear from this that marginal external costs of accidents will differ when any of these parameters differs, and that direct transfers of values from one context to another may only be done if it is believed that all these parameters are unchanged.

We expect the external marginal cost to be high if:

- the accident risk r is high;
- the cost per accident is high $(a + b + c)$;
- most of the costs fall on other user groups $(\theta \approx 0)$;

- the risk increases when the traffic increases ($E > 0$);
- or a large part of the accident cost is paid by the society at large (c).

We believe that this method is suitable for all modes in all member states. We cannot foresee any more general form of the external marginal accident cost, except that risk-avoiding behaviour should be introduced. This involves formidable practical issues in terms of estimation, however.

The basic function to discuss is the accident function. The form of this function is captured in the accident risk, r, and the risk elasticity, E.

While the various case studies do not give a clear recommendation, Chapter 6 shows how risk elasticities vary for example by road type, level crossing type and axle weight. We assume that the risk elasticity can be generalised and used in other member states. However, we have only reported a limited number of studies and some of our results are lower than expected. More case studies should be carried out before we are prepared to suggest a set of transferable and reliable elasticities.

The accident risk will define the level of the accident function. Estimates of accident risk are often easy to collect and available in each member state. However, for some modes it is difficult to estimate the risk due to lack of data. A priority has to be to improve the data availability.

The cost share that falls on a user category (θ) could possibly be generalised to other member states provided that their insurance regimes and charging systems for public services are similar. However, this information should be possible to find within country accounts. A trend function for the reduction in risk over time should be used and the function is thus general (even if it might not be linear).

The VOSL remains the subject of some controversy; the value used within UNITE is based on our reading of the best evidence possible. In common with other money values, procedures for its transfer from one context to another are discussed in Section 4. Other costs of accidents are not easily transferred, as they will depend on factors such as the mix of vehicles of different types, nature of property close to roads etc., although if these are believed to be similar it might be reasonable to make such transfers simply using purchasing power parity exchange rates as also discussed in Section 4.

3.5. Generalisation Aspects of Marginal Environmental Costs

The preferred approach to the estimation of environmental costs of transport is the impact pathway approach, which first forecasts the emissions of each pollutant, then their dispersion and finally looks at all the various impacts these have and costs them. The overall impact pathway approach is generally transferable for all categories of pollutant. However, it should be considered that in spite of significant progress made in recent years, the quantification and valuation of environmental damage still suffers from significant uncertainties. The following list summarises the crucial areas of uncertainty.

Effects of particles on human health. The dose – response models used in the analysis are based on results from epidemiological studies. However, at present it is still not known whether it is the number of particles, their mass concentration or their chemical composition which is the driving force.

Effects of nitrate aerosols on health. Nitrate aerosols are a component of particulate matter, which we know cause damage to human health. However, in contrast to sulphate aerosol (but similar to many other particulate matter compounds) there is no direct epidemiological evidence supporting the harmfulness of nitrate aerosols, which partly are neutral and soluble.

Valuation of mortality. ExternE recommends the use of *Value of a Life Year Lost* (VLYL) rather than the VOSL for the valuation of increased mortality risks from air pollution. This approach is still controversially discussed in the literature. The main problem for the VLYL approach is that up to now there is a lack of empirical studies supporting this valuation approach.

Omission of effects. Impacts on, e.g., change in biodiversity, potential effects of chronic exposure to ozone, cultural monuments, direct and indirect economic effects of change in forest productivity, fishery performance, and so forth, are omitted because they currently cannot be quantified.

The following tables show the list of input values for the estimation of air pollution and noise marginal costs and their relationship with the generalisation issues.

Concerning air pollution, input to dispersion models, due to their highly site-dependent nature, cannot be generalised to other case studies. The exposure-response functions (ERFs) for the acute and chronic health effects of air pollution, on the other hand, can be generalised (Table 1).

Concerning the generalisation of monetary values, the use of country-specific adjustments is recommended, i.e. the Purchasing Power Parity structure.

Emission factors for specific vehicle technologies (e.g. passenger car complying with EURO2 standard) can be transferred to other countries/locations. Emission factors for vehicle fleets (e.g. of a country, on a certain road) are not generally transferable, because the fleet composition usually varies.

Table 1. Input Values and Generalisation Issue for Air Pollution.

Input Values	Generalisation Aspects
Inputs to dispersion models	Generalisation not recommended
Exposure-response functions	Can be generalised
Monetary values for health impacts	Country-specific adjustment/values for local scale impacts
Exhaust emission factors for specific vehicle technologies	Same emission standard; same driving characteristics/speeds
Exhaust emission factors for vehicle fleets	Generalisation not recommended
Emission factors for the production and transport of fuel	Refinery processes and fuel distribution are comparable

The generalisation of specific exhaust emission factors for vehicle technology should be carried out taking into account the driving characteristics and the average speed. This explains why, on the other hand, the generalisation of exhaust emission factors for vehicle fleets is not recommended.

Emission factors for the estimation of up and down-stream processes can be generalised.

Aircraft emissions are a special case as most of the emissions take place in high altitudes. Assessment of the resulting impacts is still to be improved, because modelling of dispersion and chemical conversion is not as advanced as for low-level emissions. Other impacts due to low-level emissions at airports can be assessed with existing models.

Concerning marginal noise estimates, the ERFs, as for the air pollution costs, can be generalised. Monetary values should be generalised according to country-specific adjustment factors (Tables 2 and 3).

Noise emissions factors for vehicle types should be generalised taking also into account the driving characteristics and the average speed. Generalisation of noise emissions related to vehicle fleets is not recommended.

A comparison of the case study results suggests that a direct transfer of costs due to air pollution cannot be recommended. Some general rules could be derived, but an operational formula for transfer requires a broader statistical basis of case studies. A generalisation methodology for air pollution costs should account for:

- the local scale conditions (population density and local meteorology);
- the regional scale costs per tonne of pollutant emitted in a certain area (e.g. on NUTS1 level).

Table 2. Input Values and Generalisation Issue for Noise.

Input Values	Generalisation Aspects
Exposure-response functions for health impacts	Can be generalised
Monetary values for health impacts and amenity losses	Country-specific adjustment/values
Noise emission factors for vehicle/ train/ aircraft types	Only if same driving characteristics/speeds
Noise emission factors for vehicle fleets	Generalisation not recommended

Table 3. Output Values and Generalisation Issues for Air Pollution.

Type of Cost	Location	Generalisation Issues
Regional scale unit costs per tonne of pollutant	Extra-urban	Pollutant is emitted in the same geographical area (e.g. administrative unit on NUTS1 level)
Local scale unit costs per tonne of pollutant for low-level emissions from vehicles with internal combustion engine	Urban	Comparable local environment, i.e. population density and local meteorology; country-specific adjustment of monetary values
Costs due to fuel production (per litre of fuel or per vehicle kilometre)	Extra-urban	Comparable emission factors for production and transport of fuel; pollutants are emitted in the same geographical area
Costs from exhaust emissions per vehicle kilometre	Urban, extra-urban	Comparable emission factors and local environment and geographical area/regional scale unit costs

It is suggested that at the local scale (i.e. up to ca. 25 km from the emission source) the damage depends mainly on the density of population in that area. In this case, the transferability of estimates to locations could be realised as follows:

- Costs per unit emission per location type can be transferred to locations where the same economic values are to be applied.
- Physical impacts per unit emission per location type can be transferred between countries if different economic values are to be applied.
- On the regional scale, that is covering Europe, the air chemistry (which implies non-linearity) becomes important, along with the geographical location within Europe. In this case the estimates are transferable for adjacent countries, with the same environmental characteristics. To facilitate a generalisation of the damage estimates on the regional scale, unit values per country could be produced.

Marginal costs due to noise exposure are mainly determined by the following cost drivers:

- the distribution and distance of exposed persons from the source,
- the existing noise level, which in most cases is dominated by traffic (number of vehicles or trains per hour, mix of vehicle or train types, speed),
- the time of day (variation in disturbance effect of noise).

Such characteristics imply a difficult generalisation process of costs per vehicle/train kilometre or aircraft movement. Generalisation is difficult due to high sensitivity concerning local characteristics. For rough estimates of the order of magnitude though, a cautious transfer of output values may be undertaken for locations with similar characteristics and background noise levels.

For global warming, damage cost estimates (per unit emission) are globally applicable, as the location of emissions is irrelevant. Avoidance costs (per unit emission) are expected to differ between countries, as the most efficient measures to avoid greenhouse gas emissions may be different and may cause different costs, unless international emissions trading measures are in place to equalise these.

4. TRANSFER OF MONEY VALUES

As mentioned above, the monetary valuation of a given physical effect will vary between contexts and techniques exist to enable money values to be translated from one context to another. In exceptional cases, where it is only the monetary value that differs between one context and another, then this will enable the entire result to be transferred from one context to another.

More usually, translation of money values will need to be combined with translations of how the physical effects differ from one context to another if results are to be generalised or transferred between contexts. This section outlines the valuation methods used in various contexts by the UNITE project, commenting on their transferability. We consider first some particular issues regarding transfers in space and time. We then comment specifically on key parameters – the value of a statistical life, the value of a life year lost, the value of time and the discount rate.

In some cases monetary values might be expressed as 'common' to a set of countries or as 'specific' to a particular country and a choice is required as to which value to use. In general, UNITE avoided the use of common (i.e. average) European values on the grounds that they hide genuine differences in willingness to pay (WTP) between countries. Where national-level WTP-based studies exist and are consistent with the state-of-the-art, values from these studies can be adopted. Where these conditions are not fully met, the second best approach will be to take values from the state of the art WTP studies and apply benefit transfer methods to other countries. While we are aware of the equity argument in favour of using the same value of parameters such as the value of a statistical life or the value of time across all countries regardless of income, we do not consider this an appropriate way of handling differences in income between countries. To charge people in poorer countries for congestion and pollution they cause as though they were inflicting them on people of average European income appears to us to be inefficient and unfair. We prefer to take distributive considerations into account through the use of explicit equity weights.

When transferring money values between countries, it is not sufficient to use these official (nominal) exchange rates because they do not reflect differences in purchasing power, which are the best guide we have to differences in WTP in different countries. A Purchasing Power Parity (PPP) adjustment is required, specifically: multiplication of the value by the ratio of Real gross domestic product (GDP) per capita (at PPP) in the second country to Real GDP per capita (at PPP) in the first country. For countries in the Eurozone, only this PPP adjustment will be required – no exchange rate is involved. For countries outside the Eurozone, the exchange rate and PPP adjustments can be made simultaneously by using the Organisation for Economic Co-operation and Development (OECD) GDP per capita (at PPP) series. The issue of the sequence of events may arise: in general, starting with, say, 1999 data, the first step should be to deflate to the price base year (1998) using the country-level general price index for the first country, *then* apply PPP exchange rates for 1998.

The growth in real incomes is the main factor that requires adjustment to values over time. In general, we assumed that values grow with real incomes, based on an elasticity of 1.0. After much discussion of the rather limited evidence available we have applied this to all WTP values in UNITE, and believe this to be the best approach to transferring values to use in general unless there is robust evidence to the contrary.

There is a reasonably widespread agreement that monetary values of risk reductions in the transport sector should be defined so that they reflect individual preferences of the affected population. The value should be expressed as the affected individuals collective WTP for safety improvements or willingness to accept compensation (WTA) for increased risk. This can be estimated by asking a sample of the affected population about the amount they would be willing to pay or accept as compensation for changes in the level of safety, a method which is often referred to as the 'contingent valuation' (CV) method. From the WTP a 'value of statistical life' 'VOSL' can be derived.

In UNITE we proposed a European Standard Risk Value based on a limited number of well-designed CV studies. We used the value 1.5 million Euros (M€)per fatality measured in market prices. To have the full value of a fatality the cost of net lost production, medical and ambulance costs should be added, which is approximately 10% of the risk value. To express it as a factor cost, it should be reduced with the proportion of indirect taxation (approximately 20%). As a sensitivity test a higher VOSL of 2.5 M€ is used as it is at the upper end of reliable state-of-the-art studies. If also a low value was tested, a value of 0.75 M€ was used.

For individual countries we recommend the following procedure. If a national value exists, if it is based on the WTP/WTA principle and if the basic study is well designed it is used. In the absence of national values, a European Standard Risk Value should be used, adjusted in accordance with real per capita income at purchasing power parity exchange rates for each country.

The quantification of health risks due to air pollution in UNITE is based on ERFs, giving a relationship between ambient pollutant concentrations and health effects. To quantify the changes in mortality, the population affected by pollutant exposure has to be tracked over time, because the effects show in later years. This means that based on Pope et al. (1995), we can quantify lost life years and not the so-called premature deaths. To account for this, the YOLL (Years of Life Lost) approach was adopted in ExternE, implying a valuation of lost life years and not fatality cases. So in the first place the use of the VLYL follows the requirement of the underlying study.

In the absence of empirical data on *VLYL*, the *VLYL* can be estimated from the value of statistical life (VSL) according to the following relationship:

$$VSL_a = VLYL(r)\sum_{i=a}^{T}\frac{P_{ai}}{(1+r)^{i-a}}$$

where:

VSL_a is the value of statistical life;
a the age of person whose VSL is being estimated;
$VLYL(r)$ value of a life year lost;
r the discount rate;
T maximum life expectancy (100 years in our calculations);
P_{ni} (conditional) probability that a person of age n will reach age i.

Uncertainties remain considerable in the monetary valuation of environmental mortality risks. There is, for example, evidence suggesting that WTP for reducing environmental mortality risks is higher than for traffic accident risks. Jones-Lee, Loomes, Rowlatte, Spackman, and Jones (1998) propose a factor of two to transfer the VSL for road accidents to the air pollution context. In the case of UNITE this means, that the VSL for road accidents is multiplied by two to account for the environmental context. Based on this value the VLYL is then derived according to the equation above. Starting from a value of 1.5 M€ (as recommended) this implies a VLYL of 95,000 and 150,000 € for discount rates of 0% and 3%, respectively. These values are applicable for acute effects, i.e. effects which appear immediately after pollutant exposure. For chronic effects, i.e. effects which occur with delay, the time lag between pollutant exposure and occurrence of effect has to be taken into account and discounted accordingly. Transfer between countries follows the same approach as for the value of a statistical life.

The best basis for values of time (VOT) is to take them from a limited set of state-of-the-art research studies using a consistent methodology; studies such as the Dutch national value of time studies 1986–1995 (Hague Consulting Group, 1996), the U.K. value of time study 1994 (Hague Consulting Group, 1994), the Swedish national value of time study (Alger, Dillen, & Widlert, 1996), for freight, the work carried out by De Jong and the Hague Consulting Group in 1993 for the Netherlands (De Jong, 1996), and for values under different travel conditions in passenger transport (Wardman, 1998). The second choice basis is to transfer values from these state-of-the-art studies to other settings.

For passenger travel the following differentiation of the values of travel time should be made where the data permits:

- Travel purpose (business, commuting, leisure);
- Mode (car, bus, rail, air);
- Travel distance (urban/local, inter-urban/long distance);
- Travel condition (expected travel time, delay time, in-vehicle/walk/wait).

In general the empirical results cannot support the level of disaggregation, and instead the values are split simply into:

- Road (values per vehicle hour):
- Light goods vehicles
- Heavy goods vehicles

Other modes (values per ton hour)

- Rail freight
- Inland waterway
- Maritime
- Air freight

For delay time, the evidence would seem to support 1.5 as a conservative estimate of the factor on expected in vehicle time. Based on the analysis in Hague Consulting Group (1996), in-vehicle time values should be multiplied by a factor of 1.6 if walking or waiting values are required.

In order to apply the UNITE values in specific countries, the values should be adjusted in accordance with real per capita income at purchasing power parity exchange rates for each country. To reach a final UNITE value for commuting and leisure time, VOT should be divided by $(1+\tau)$ (where τ is average indirect taxation) for the country concerned.

When estimating marginal social cost, infrastructure costs should be valued on a social basis, so when discount rates are required they should be *social* discount rates. There are various possible bases, discussed at length in Lind (1982). Generally, we feel that a social opportunity cost rate is appropriate. Furthermore, all prices in UNITE are constant 1998 prices, so the discount rate should be a real rather than a nominal rate. In view of the evidence, the standard rate of discount used in UNITE was a 3% real rate. Although this might in principle differ between countries, we would expect that within Europe there would be sufficient mobility of capital that a similar rate would be appropriate for all countries.

5. CONCLUSIONS

In this section we summarise our conclusions about the extent to which marginal cost estimates may be adapted or transferred from one context to another. The transfer of marginal cost studies has been viewed in terms of:

- Transfer of methodology
- Transfer of functional relationships and input values
- Transfer of output values

We consider each of these forms of transferability in turn:

5.1. Transfer of Methodology

In general, best practice methodologies can be transformed between contexts for instance, for accident costs and the Mohring effects, single formulae are presented which may be transferred between contexts. However, there are problems with simply recommending transfer of methodology. In the case of infrastructure and supplier operating costs, an econometric approach is preferred, but data availability is a big problem, and even where data are available it is rarely possible to obtain results at the level of disaggregation needed for pricing purposes (e.g. by vehicle type). Thus econometric methods need to be supplemented by engineering and cost allocation methods, perhaps transferring cost elasticities from econometric studies elsewhere. Similarly, the preferred bottom-up studies of congestion costs and environmental costs are expensive, so simpler approaches to transferability would be desirable.

5.2. Transfer of Functional Form and Input Values

Some relationships, such as speed/flow relationships, have been extensively studied and may be transferred to similar locations elsewhere. Similarly, as explained above, parameters such as cost elasticities and accident risk elasticities may be transferred to similar situations elsewhere. In general economic values, such as VOT and VOSL may be transferred, adjusting for real incomes and purchasing power parities. But most other relationships and input values have to be freshly estimated for each context.

5.3. Transfer of Output Values

The earlier chapters and analysis of the transferability possibilities undertaken in this chapter demonstrate, in the first place, that the availability of marginal costs estimates is uneven across the analysed cost categories. Road and rail, on the one hand, are widely documented thanks to a substantial number of case studies on a significant range of cost categories, i.e. environmental costs, infrastructure costs, transport users costs. On the other hand, aviation, inland waterways and maritime transport have only been analysed in a fairly limited number of case studies.

This reflects heavily on the potential comparability of results. In fact, the meaningfulness of comparisons across sites and situations is directly geared to the size and quality of the sample of original observations. The higher the number of available case studies, the better the chance of identifying meaningful relationships between marginal cost values and the corresponding cost drivers.

In order to ensure the comparability of marginal costs estimates, all of the following minimum requirements must be met:

- a common measure for monetary values at a given time (year) : VOT for transport user marginal costs, VOSL for estimating accidents and environmental costs etc. should be related to a specific year, using purchasing power parity for transferring values among countries and adjusting for inflation if a different year is available;
- a common methodological background: for instance, with reference to the marginal accident costs, many studies do not adequately distinguish external from internal costs;
- a common set of cost drivers: to ensure, in turn, a common framework of results, e.g. environmental marginal costs for road traffic must be differentiated along the same vehicle classification etc.

In particular cases, (e.g. supplier operating costs, air pollution), the possibility to use average costs as an approximation to marginal costs should be taken into account.

The quantification of marginal costs is a complex task and, in some cases, as for environmental marginal costs and infrastructure costs, it is a time-consuming process, requiring in-depth analyses for the identification and documentation of the cost functions. Devising a robust methodology for transferring results across sites and situations could therefore yield substantial benefits to a wide range of potential users.

It must be stressed that the above transferability process cannot be implemented with the same degree of confidence throughout the external cost categories. Although much of the effective implementation depends on sheer data availability, the level of difficulty in implementing the transferability process varies with the cost categories, as summarised below.

External Cost Category	Difficulty in the Implementation of the Transferability Process
Wear and tear	Low
Congestion	Low
Accidents	High
Air pollution	High
Global warming	Low
Noise	High
Up- and down-stream	Low

Quite obviously, one can observe that the more a given external cost category relies on bottom-up data, the higher will the level of difficulty in ensuring a reliable transferability process be.

As with the basic requirements for comparability of results outlined above, one can conclude that the transferability of results should be based on a similar set of basic pre-conditions.

(1) A common methodology for estimating marginal costs should be adopted at the outset, in order to minimise the problem of comparability.
(2) A large number of marginal cost (reference) values should be available, in order to guarantee the robustness of the statistical analysis, i.e. regression analysis, carried out to determine the adjustments necessary to transfer results from one context to another.

Finally, extreme caution should be used for the transferability of the specific marginal costs category that is highly site-dependent, with particular emphasis on congestion costs.

We believe there is a strong argument for undertaking detailed case studies where the data, expertise and time are available so as to generate robust, up-to-date and reliable estimates, and then using these case study findings as the basis for estimate where the data, expertise and time are not available. Given the difficulties referred to above, the idea of transferring functional forms, input values or output values from one context to another is attractive. We have shown that in some cases this may be a reasonable approach. However great caution is needed, from some categories of cost there is great variability with context and not enough case studies have been done to understand fully this variability. Furthermore, evidence from the IMPRINT-EUROPE thematic network project revealed a demand for the introduction of standard data collection formats and procedures and for simplified practical tools for cost estimation to support price setting as a matter of urgency. Robust methodology for cost and impacts calculation are, it would appear, not in themselves a guarantee of a successful implementation if the gap with the policy world is not correctly bridged. Thus, in general more state-of-the-art case studies are still needed and further guidance on generalising cost estimates and mechanisms for deriving these are required. These are key features of the GRACE project, the follow-up project to UNITE.

REFERENCES

Alger, S., Dillen, J.L., & Widlert, S. (1996). The national Swedish value of time study. Paper presented at the PTRC International Conference on the Value of Time.

De Jong (1996). Freight and coach VOT studies. Paper presented at the PTRC Seminar on the Value of Time, Wokingham, UK.

Gaudry, M., & Quinet, E. (2003). *Rail track wear-and-tear costs by traffic class in France*. Universite de Montreal, Publication AJD-66.

Hague Consulting Group. (1994). *UK Value of Time Study*.

Hague Consulting Group. (1996). The 1985–1996 Dutch VOT studies. Paper presented at the PTRC Seminar on the Value of Time, Wokingham, UK.

Jones-Lee, M., Loomes, G., Rowlatte, P., Spackman, M., & Jones, S. (1998). *Valuation of deaths from air pollution*. Report for the Department of Environment, Transport and the Regions and the Department of Trade and Industry, London.

Lind (1982). Discounting for time and risk in energy policy.

Lindberg, G. (2002). Marginal costs of road maintenance for heavy goods vehicles on Swedish roads. *UNITE (UNIfication of accounts and marginal costs for Transport Efficiency) Deliverable 10, Annex A2*. Funded by EU 5th Framework RTD Programme. University of Leeds, Leeds: ITS. http://www.its.leeds.ac.uk/projects/unite/

Pope, C.A. III, Thun, M.J., Namboodiri, M.M., Dockery, D.W., Evans, J.S., Speizer, F.E., & Heath, C.W. Jr. (1995). Particulate air pollution as predictor of mortality in a prospective study of US adults. *American Journal of Respiratory and Critical Care Medicine*, 151, 669–674. *Vergelijkingskader modaliteiten*. Rijswijk: NEA, Sterc, Transcare.

Sansom, T., Nash, C., Mackie, P., Shires, J., & Watkiss, P. (2001). Surface transport costs and charges: Great Britain 1998. Institute for Transport Studies, University of Leeds in association with AEA Technology Environment.

Wardman, M. (1998). The value of travel time: A review of British evidence. *Journal of Transport Economics and Policy*.

POLICY CONCLUSIONS

Chris Nash and Bryan Matthews

1. INTRODUCTION

It is clear that measurement of short-run marginal social cost is difficult. Despite them having been much studied, there is still considerable uncertainty as to the true marginal social cost of infrastructure wear and tear and of congestion. In contrast, there is little research into scarcity costs, which arise on those modes where use of the infrastructure is scheduled and the presence of operators filling all the slots make it impossible for anyone else to get access to the infrastructure at the time in question. Enormous progress has been made on the measurement and valuation of environmental costs and external accident costs in recent times, though there are still uncertainties in these areas too.

However, much progress on cost estimation has been made over recent years, thanks in large part to EU research projects such as EXTERNE, QUITS, RECORD-IT and UNITE, and the inability to measure marginal social cost cannot now be held to be a major barrier to implementation. Although inevitably problems and disagreements remain, there is generally enough information available to permit the setting of prices that are more closely linked to marginal social cost than they are at present, and which provide for a more efficient use of the transport system, even if such prices do not represent the optimum. It has been said on several occasions during the IMPRINT-EUROPE seminars, which brought together policy and research stakeholders to discuss the implications of transport pricing-related research findings, that 'we should not let the best be the enemy of the good'.

Measuring the Marginal Social Cost of Transport
Research in Transportation Economics, Volume 14, 315–326
Copyright © 2005 Published by Elsevier Ltd.
ISSN: 0739-8859/doi:10.1016/S0739-8859(05)14011-6

UNITE considered four types of infrastructure-related costs, as well as transport supplier operating costs:

- use-related wear and tear costs;
- congestion and scarcity costs;
- external accident costs; and
- environmental costs.

We concluded that infrastructure wear and tear cost is best measured by an allocation process informed by econometric studies. There is a good degree of consensus as to what are the key variables determining costs, though disagreement remains as to the precise nature of the relationships so an element of judgement is involved.

Similarly for supplier operating costs, we believe that the cost allocation approach will continue to be needed for pricing purposes, but that the allocations should be informed by the results of econometric research. Estimation of 'price-relevant' costs here also appear to depend on whether the operator reacts to changes in traffic by changing vehicle size or train length or by changing the frequency or pattern of services; in the latter case it would seem necessary to compute both supplier operating costs and the Mohring Effect, i.e. the user benefit arising from the change in service level.

There is an established approach to measure congestion costs, though it is not clear whether the highly variable results which emerge in this area are related to different modelling techniques or to actual differences in circumstances. There are also concerns about data availability and the lack of studies on modes other than road. The extent to which pricing reform can contribute to the efficient allocation of scarce capacity in rail and air remains uncertain because of the complexity of cost measurement and implementation in pricing policies.

Identifying the external component of accident cost remains uncertain because of the limited amount of evidence on risk elasticities. Traditionally it has been thought that extra traffic will make it less safe for those already on the roads, but it is not clear that this is always the case. There has been evidence that in some cases extra traffic that adds to congestion actually makes roads safer, so there is still work to do. In the meantime, it is reasonable to regard external accident costs as relatively low, as was found in the case studies in the UNITE project.

Great progress has been made on measurement and valuation of environmental cost, in particular noise and air pollution, though uncertainties and disagreements remain, particularly regarding the valuation of global warming. A common methodological framework – the impact pathway approach – has been applied in almost ten different countries to derive estimates for all modes of transport. There are a large number of input functions used in the approach and concerns have been expressed about the transferability of some of these functions from one set of conditions to another. Hence, further development of local knowledge, in particular relating to how functions apply in different conditions, is necessary.

Hence, most countries have some information suitable for setting more efficient charges than those which currently exist, though the disagreements and uncertainties identified above serve as a barrier to estimation of charges according to a common basis, consistent with promoting equal terms of competition. Nevertheless, research is rapidly reducing the uncertainties, and 'the use of proper theory and modern methods will lead to a convergence also of the more difficult marginal cost categories in the near future' (Lindberg, 2002). In other words there is no reason for measurement problems to hold up moves towards marginal social cost pricing. In any event it is hard to argue that, were marginal social cost the right concept to use in pricing, measuring something else instead of using the best estimate possible would be a sensible approach. A likely way forward is to seek consensus on a lower limit of costs that should be reflected in price, in tandem with any upper limit that might be placed on prices for reasons of public and/or political acceptability; these limits could then rise as evidence becomes stronger and more accepted, and as public and political acceptability concerns subside. We believe that the values used in the UNITE project and put forward in this volume, represent such a lower bound. Other studies (e.g. IWW and INFRAS, 2004) which include a wider range of effects, use different dose–response functions and use more demanding targets (e.g. for global warming) produce much higher values but on the basis of more uncertain evidence.

We have argued that there is often 'good enough' information to enable policy-makers to proceed with implementation of pricing policies based on marginal social cost and, have, in the previous chapter, reviewed the range of estimates which exist. We now turn to re-examine the benefits of using marginal social cost as the basis of transport pricing, before discussing some of the issues that pose barriers to the use of marginal social cost based pricing in practice.

2. TOWARDS IMPLEMENTATION OF MARGINAL SOCIAL COST BASED PRICING IN TRANSPORT

Early on in the IMPRINT-EUROPE thematic network project it became clear that considerable divergence of views on pricing doctrines, used to form the basis of policy-making, and of practices exists in different countries throughout Europe, reflecting in part different existing circumstances and objectives, e.g. between congested core members and peripheral regions, environmentally sensitive areas and others. In Chapter 2, Quinet reviews the different pricing doctrines in use throughout Europe and identifies three distinct doctrines being expressed by political authorities in different countries, one broadly in line with marginal social cost pricing (though tending to favour long run rather than short run marginal social cost), and two based on average cost pricing (one stemming from concerns about the overall government budget and the other stemming from particular concerns about the transport budget). Understanding the nature of these different pricing doctrines and practices, the ways in which they differ and how they might be reconciled, would lift a significant barrier to the implementation of a consistent, comparable transport pricing policy across the EU.

An important part of the UNITE project involved modelling the implications of adopting alternative pricing rules. It is sometimes argued that accounts information should be used to set prices to cover total cost on each mode, perhaps on the grounds of budget constraints or that this is the most equitable way to cover the costs of the transport system. The UNITE integration work modelled the consequences of this and compared them with two other policies; pure marginal social cost pricing, and social welfare maximisation subject to a budget constraint (Ramsey pricing). Two types of modes – partial equilibrium and general equilibrium models were used.

When people are confronted with transport accounts, a common reaction is that costs and revenues should be balanced, which implies a form of average cost pricing. In UNITE Deliverable 4 (Mayeres, Proost, Quinet, Schwartz, & Sessa, 2001), a conceptual analysis of the use of transport accounts for pricing showed that transport accounts are a useful source of information for pricing policies but should not serve as guideline or criterion for transport pricing. In Chapter 8, Proost and Mayeres analyse this issue further by quantifying the welfare effects of average cost pricing. As expected it is found to be substantially inferior to marginal social cost pricing, and indeed in the case studies presented inferior also to the status quo. The argument sometimes heard that average cost pricing is fairer than marginal social cost pricing is also discredited by these case studies; it is found that all income groups are made worse off by average cost pricing, with the greatest proportionate loss resting on the poorest.

It is clear that EU-wide political consensus on the need for transport pricing reform is growing and our evidence from the IMPRINT-EUROPE thematic network project showed that policy-makers are actually more interested in the impacts and benefit calculations of different pricing reforms than in theoretical justifications of pricing principles. The forum of ECMT has done much to facilitate political consensus-building. Two ECMT resolutions on the subject have been adopted in recent years by transport ministers. Resolution 1998/1 on the Policy Approach to Internalising the External Costs of Transport (ECMT, 1998); and Resolution 2000/3 on Charges and Taxes in Transport and Particularly International Road Haulage (ECMT, 2000). These resolutions promote a gradual, stepwise reform of charges and taxes to improve the efficiency of transport, avoid discrimination and distortion of competition and provide incentives to reduce the environmental impacts of transport and manage congestion. At the 2000 ECMT Council meeting, transport Ministers agreed "to aim at gradually shifting the structure of taxation in transport to increase the share of more territorially based taxes and charges (e.g. tolls and km-charges) as this contributes at the same time to: ensuring non-discrimination; improving efficiency (and Ministers noted that marginal social costs are the most efficient basis for formulating charges); avoiding problems of competitiveness between national haulage industries; and promoting sustainability". Most recently at the 2003 ECMT Council, transport Ministers approved a report on Reforming Transport Taxes that concludes (ECMT, 2003):

(1) The potential benefits of the reforms set out in the resolutions are large.
(2) There are no arguments of principle that give reason to delay reform.
(3) Focus is now on implementation and carrying public opinion.

Whilst the policy prescription to base prices on marginal social costs is relatively simple in principle, in reality complexities inevitably crowd in. As we argued in the opening chapter, however, this does not mean that a totally different theoretical approach to pricing policy needs to be adopted, or that full cost recovery as a principle is a good starting point. For implementation, the questions then turn to what to do in the face of these complexities and what further practical points can be agreed even amongst those who disagree about the principles?

It is widely recognised that various barriers place constraints on the potential for implementation of full marginal social cost pricing. The MC-ICAM project (Nash, Niskanen, & Verhoef, 2003) made a conceptual distinction between underlying barriers and implied constraints to which they led. Factors that may lead the barriers and constraints to be eased over time were also considered.

The best way to devise infrastructure prices was seen as identifying the barriers that prevented the full implementation of marginal social cost pricing and the constraints to which they led. Knowing these constraints, the second-best optima could be found. The next stage is to identify the succession of second-best optima, each better than the last, which becomes feasible after the constraints are eased. This set of second-best optima is essentially the implementation path. If all barriers can eventually be overcome then the implementation path leads eventually to full marginal social cost pricing; if not, it leads to the best 'second best' solution.

The IMPRINT-EUROPE thematic network project considered a wide range of experience from the implementation of pricing reforms in many countries, including urban road pricing in Norway and London, heavy goods vehicle charges in Switzerland and rail infrastructure charges in Britain, Sweden and Austria, as well as the results of many modelling case studies and demonstration projects. One thing that quickly became clear was that existing transport pricing practices vary considerably between countries and, to an even greater extent, between the different modes. So we cannot think of a simple movement towards marginal social cost based pricing, where all prices in all countries move in the same direction; the implementation path will be very different from place to place and from mode to mode.

MC-ICAM identified separately the barriers and constraints for urban and interurban road, rail, air and sea. The three types of barriers to implementation were technological, institutional, and public and political acceptability. Conclusions on barriers were as follows

Technological barriers:

- The big issues in terms of urban and interurban road transport are cost and reliability of the technology and confidence that it will work rather than availability per se.
- Inter-operability is seen as important especially in terms of interurban road pricing, where different countries are going in different directions and inter-operability is seen as crucial to the avoidance of waste.
- For urban public transport, smart card technology was removing the technological barriers to full marginal social cost pricing.
- The big issue for rail, air and sea is not the technology but creating appropriate ways of measuring the costs of congestion and scarcity of capacity and reflecting these in charges. Further work on measuring cost of scarcity is underway in projects such as SPECTRUM.

Institutional barriers:

- There is the need for EU and national legislation to permit and support marginal social cost pricing.

- The relationship between different levels of government leads to institutional problems. Marginal social cost pricing may be best implemented if all decisions were centralised and taken by the state or EU but there is a risk of government failure.
- Issues arise in the roles of deregulation and privatisation versus government control when implementing marginal social cost pricing. An approach would be to deregulate and privatise firms because competitive markets would force firms to implement marginal cost pricing to survive, and externalities could be accounted for by the use of Pigovian taxes and subsidies. However, due to the market power some firms possess especially in cases of natural monopolies, firms may not apply marginal social cost pricing; therefore, government intervention may be needed. Government control may risk government failure, as governments may not have enough information or motive to act in the public's best interest.

Acceptability barriers:

- Public and political acceptability is the key barrier.
- A number of points were seen as important in order to gain acceptability. It was best to start off with a simple system (e.g. cordon charges) and move to a more complicated system as confidence builds up. Packages and measures within and across all modes such as environmental charges on all modes at once help acceptability. There is the need for increases in charges to be gradual and the way the revenue generated is used is crucial.
- Many factors make acceptability less of a barrier in interurban road transport than in urban road transport. These include more complexity in urban networks and the fact that existing taxes are seen as unfair since they focus on where a vehicle is registered and fuelled rather than where it is running, which is common in interurban road transport.

The use of revenues is found to be a crucial issue in transport pricing, both in terms of its link to public and political acceptability, and in terms of its impact on the welfare effects of pricing reforms. In general, one might expect that effective use of pricing revenues would re-enforce the positive welfare effects of transport pricing reform. Importantly, the MC-ICAM case studies found that whilst constraining the use of revenue to be returned to users through other tax reductions or to be used in the transport sector may well increase acceptability, it actually reduced the benefits of pricing reform.

In summary, acceptability tends to be higher where problems are particularly acute and demonstrable, where revenue use is transparent and/or earmarked and where there is an identified package of complementary measures, whilst there is evidence to show that acceptability is higher where initial price changes are simple and modest and that acceptability for more sophisticated charging may grow over time.

The barriers and constraints referred to above give rise to issues concerning the speed of implementation, which may differ between the modes, essentially because the technical requirements differ and for political reasons. In turn, these factors lead to second-best considerations; differing degrees and speed of implementation between the modes could itself be distorting and needs to be taken into account in policy determination. In addition, implementation of sophisticated pricing systems remains far from costless, whilst the ability of users to respond to the prices may be limited by their ability to understand and predict what they will have to pay. Thus there is an optimal degree of complexity of any price structure arrived at by comparing the benefits of greater differentiation in terms of their influence on the volume and location of traffic with the costs.

Phasing was a particular issue examined by the MC-ICAM project, which defined implementation paths in terms of sequences of consecutive second-best optima. It was found that second-best optimisation may lead to initial price levels that have to be changed radically later when constraints are changed or relaxed. The finding that second-best pricing along an implementation path may cause certain prices to fluctuate and problematic reverses in the direction of movements of charges may give rise to a caveat. Second-best optimisation that allows for current distortions in other modes or markets may lead to problems in implementing policy in the longer term which may lead to issues with acceptability. An important lesson from MC-ICAM was that 'rather than focussing on fine-tuning of the derivation of second-best prices, and often very detailed technical problems related to implementation, the policy-makers and analysts should pay more attention to identifying the key barriers and their implied constraints, and how to avoid or remove them'.

Thus implementation of marginal social cost pricing is far from straightforward, but there seems to be good evidence that simple systems of road pricing in heavily congested cities or on congested interurban links can bring worthwhile benefits (Nash et al., 2003), whilst more complex systems are becoming steadily easier to implement as technology advances. So making simple and modest reforms first, progressing towards more sophisticated charging systems, can address concerns about reform, whilst each new phase of reform should generally move prices in the right direction in terms of the ultimate goal, even if short-term considerations might dictate otherwise.

In summary, then, whilst it is best wherever possible to go straight to the first-best solution, there were also good scientific and practical reasons for the need for phasing and packaging. Some barriers definitely stood in the way of achieving first-best solutions, therefore, second-best solutions were needed. Uncertainty tended to remain about the barriers and the exact nature of the constraints to which they lead to and the degree to which they could be eased over time, including the level of government intervention. But the MC-ICAM methodology is seen as a valuable step forward.

3. CONCLUSIONS

The UNITE case studies have produced estimates for a wide range of circumstances and using a wide variety of approaches. In some cases (such as supplier operating costs for rail and air), these have been the subject of much previous analysis; in others, such as infrastructure costs for water transport, the existing literature is very sparse.

These case studies illustrate that there is no unique 'state-of-the-art' approach for the estimation of marginal costs. For instance, for infrastructure and supplier operating costs, there may be a preference for econometric estimation, but it is seldom possible to do that at a level of detail that lends itself directly to pricing decisions; a mixture of econometric research with engineering and cost accounting approaches, therefore, is necessary. For user costs, road congestion has been extensively investigated using either single link or network models based on speed flow relationships and junction delay formula. For rail and air congestion, a regression-based approach has also been developed but work in this area is new; there are few existing studies. No estimates of scarcity costs for these modes have been found and this is a priority for future work.

For accidents, a correct methodology has been developed which requires inputs from a variety of sources; the most difficult being risk elasticities. The impact pathway approach is recommended for environmental costs, but further research on transferability of results would be worthwhile.

The quantitative results suggest that marginal infrastructure costs are generally low; it is supplier operating costs, congestion costs, the Mohring effect and in some circumstances elements of environmental costs, particularly noise, that are the most important categories.

In terms of external costs, for the car, as expected, generally congestion costs dominates, followed in the case of urban areas only, by noise and air pollution (especially for diesel cars) with global warming generally much smaller. Accident costs are also significant in urban areas. In general, the differences between areas are quite large. Outside urban areas, noise and air pollution costs are generally much smaller.

When load factors are taken into account, external costs of air and rail are generally much smaller than car. The exception to this is congestion costs for air, where our case studies, as was noted earlier, yield lower results for cars than in other comparable car studies. The congestion costs in our air case study, Madrid airport, appear to be of the same order of magnitude as for car when allowing for load factors, although the results for rail appear to be somewhat lower. We have no quantification of pure scarcity costs, which can be important for rail and air.

The UNITE project has served to illustrate that a variety of methods are needed in order to estimate marginal social cost, working together to overcome difficulties in the availability of appropriate data. A pragmatic approach, using a combination of cost allocation, econometric and engineering models is needed, with the precise approach differing between cost categories. The most important category of external cost in general for the transport sector, is congestion (especially for road transport in urban areas), although various elements of environmental cost can be important, particularly in urban areas. Both congestion and environmental costs vary greatly from context to context, so great caution needs to be exercised in transferring results from one context to another. External accident costs appear to be less important than was previously believed, mainly because earlier studies had failed to identify correctly the external element of these costs.

A general warning should be given that many earlier studies of all cost categories used methodologies that would now be considered inappropriate and their results should be used with great care. The main priorities for further research are the treatment of congestion and scarcity in rail and air transport, and the development of better methods for transferring results from one context to another.

Whilst further research as always is needed, we believe that the evidence in UNITE and other studies provides both the necessary information and a clear case for moving towards pricing based on marginal social cost on all modes of transport. We accept that many constraints will prevent the first-best situation of pure marginal social cost pricing from being achieved, but believe that there is good evidence that pricing based on marginal social cost but allowing for specific mark-ups and other deviations on second-best grounds would yield substantial benefits.

We believe that UNITE made great progress in identifying appropriate methodology and providing new empirical evidence for the measurement of marginal social cost, in putting together consistent measurements of the costs and revenues of the different modes of transport for all the countries of the EU plus Switzerland, Hungary and Estonia and in modelling broader socio-economic impacts of transport pricing reform.

Nevertheless a number of uncertainties remained at the end of the UNITE project and these form the focus for further research. Firstly and somewhat surprisingly, we found no consensus amongst researchers or policy-makers on the division of infrastructure costs into fixed and variable elements, and although highly innovative case studies undertaken as part of UNITE helped our understanding, they did not fully resolve the problem. Secondly, large variability in the estimates of road congestion costs and the costs of noise raised issues as to the degree to which this was the result of differing approaches to modelling these costs as well as true variation in time and space. Thirdly, relatively little previous work had been undertaken on air and waterborne transport, and whilst a start was made in UNITE, lack of adequate data or prior research made it difficult to draw firm conclusions. Finally, the importance of scarcity of rail capacity had not been fully understood at the time of the development of the UNITE proposal and – except for some interesting work on auctioning of capacity – it remains the case that this is largely a new area of work.

It is recognised here that the application of marginal social cost pricing is far from straightforward. Many people working in the transport sector – be they policy-makers, stakeholders or academics – remain critical of marginal cost pricing and sceptical about its potential benefits, often being more persuaded by other pricing doctrines. whilst we challenge some of the criticisms of the theory, many such criticisms do have to be acknowledged and explicitly addressed when seeking to apply the theory in the transport sector. The area has been the subject of a great deal of research and it is our belief that many of the criticisms can be adequately addressed and allowed for and that it is perhaps due to economists' failure to properly and persuasively communicate this, that remains a stumbling block.

With this book we aim to challenge the assertion, used as a key criticism of marginal social cost pricing, that marginal social costs cannot be measured, or at least cannot be measured sufficiently well. We believe that Chapters 3–7 of this book convincingly demonstrate that it is possible to measure marginal social cost in transport and that, indeed, cost estimates already exist for each cost component in different contexts. That is not to say that all of the methodological issues are fully resolved, or that we have all of the estimates we need in order to proceed with the comprehensive implementation of marginal cost pricing. There are still uncertainties and gaps in our knowledge. However, further work in this area is likely to reduce these uncertainties and close these gaps. In any case, we believe that the existing knowledge, as set out in this book, is sufficient to enable a better pricing system to be implemented than what currently exists in most places. Furthermore, there is strong evidence that the impacts of moves to base transport prices ever more closely on those marginal cost estimates are significant and positive.

REFERENCES

European Conference of Ministers of Transport. (1998). *Resolution on the policy approach to internalsing the external costs of Transport.* Paris

European Conference of Ministers of Transport. (2000). *Resolution no. 00/3 on charges and taxes in transport, particularly in international road haulage.* Paris.

European Conference of Ministers of Transport. (2003). *Reforming transport taxes.* Paris.

IWW and INFRAS. (2004). *External costs of transport update study.* CER, Brussels.

Lindberg, G. (2002). Recent progress in the measurement of external costs and implications for transport pricing reforms. Paper presented at the 2nd Imprint-Europe seminar, Brussels.

Mayeres, I., Proost, S., Quinet, E., Schwartz, D., & Sessa, C. (Eds). (2001) *Alternative frameworks for the integration of marginal costs and transport accounts* (UNITE Deliverable 4). University of Leeds, ITS.

Nash, C., Niskanen, E., & Verhoef, E. (2003). Policy conclusions from MC-ICAM. Paper presented at the 4th Imprint-Europe seminar, Leuven.

AUTHOR INDEX

SUBJECT INDEX

Printed and bound by CPI Group (UK) Ltd, Croydon, CR0 4YY
12/05/2025
01867053-0001